Practical Guide to Pressure Vessel Manufacturing

MECHANICAL ENGINEERING
A Series of Textbooks and Reference Books

Founding Editor

L. L. Faulkner

Columbus Division, Battelle Memorial Institute
and Department of Mechanical Engineering
The Ohio State University
Columbus, Ohio

1. *Spring Designer's Handbook*, Harold Carlson
2. *Computer-Aided Graphics and Design*, Daniel L. Ryan
3. *Lubrication Fundamentals*, J. George Wills
4. *Solar Engineering for Domestic Buildings*, William A. Himmelman
5. *Applied Engineering Mechanics: Statics and Dynamics*, G. Boothroyd and C. Poli
6. *Centrifugal Pump Clinic*, Igor J. Karassik
7. *Computer-Aided Kinetics for Machine Design*, Daniel L. Ryan
8. *Plastics Products Design Handbook, Part A: Materials and Components; Part B: Processes and Design for Processes*, edited by Edward Miller
9. *Turbomachinery: Basic Theory and Applications*, Earl Logan, Jr.
10. *Vibrations of Shells and Plates*, Werner Soedel
11. *Flat and Corrugated Diaphragm Design Handbook*, Mario Di Giovanni
12. *Practical Stress Analysis in Engineering Design*, Alexander Blake
13. *An Introduction to the Design and Behavior of Bolted Joints*, John H. Bickford
14. *Optimal Engineering Design: Principles and Applications*, James N. Siddall
15. *Spring Manufacturing Handbook*, Harold Carlson
16. *Industrial Noise Control: Fundamentals and Applications*, edited by Lewis H. Bell
17. *Gears and Their Vibration: A Basic Approach to Understanding Gear Noise*, J. Derek Smith
18. *Chains for Power Transmission and Material Handling: Design and Applications Handbook*, American Chain Association
19. *Corrosion and Corrosion Protection Handbook*, edited by Philip A. Schweitzer
20. *Gear Drive Systems: Design and Application*, Peter Lynwander
21. *Controlling In-Plant Airborne Contaminants: Systems Design and Calculations*, John D. Constance
22. *CAD/CAM Systems Planning and Implementation*, Charles S. Knox
23. *Probabilistic Engineering Design: Principles and Applications*, James N. Siddall
24. *Traction Drives: Selection and Application*, Frederick W. Heilich III and Eugene E. Shube
25. *Finite Element Methods: An Introduction*, Ronald L. Huston and Chris E. Passerello

26. *Mechanical Fastening of Plastics: An Engineering Handbook*, Brayton Lincoln, Kenneth J. Gomes, and James F. Braden
27. *Lubrication in Practice: Second Edition*, edited by W. S. Robertson
28. *Principles of Automated Drafting*, Daniel L. Ryan
29. *Practical Seal Design*, edited by Leonard J. Martini
30. *Engineering Documentation for CAD/CAM Applications*, Charles S. Knox
31. *Design Dimensioning with Computer Graphics Applications*, Jerome C. Lange
32. *Mechanism Analysis: Simplified Graphical and Analytical Techniques*, Lyndon O. Barton
33. *CAD/CAM Systems: Justification, Implementation, Productivity Measurement*, Edward J. Preston, George W. Crawford, and Mark E. Coticchia
34. *Steam Plant Calculations Manual*, V. Ganapathy
35. *Design Assurance for Engineers and Managers*, John A. Burgess
36. *Heat Transfer Fluids and Systems for Process and Energy Applications*, Jasbir Singh
37. *Potential Flows: Computer Graphic Solutions*, Robert H. Kirchhoff
38. *Computer-Aided Graphics and Design: Second Edition*, Daniel L. Ryan
39. *Electronically Controlled Proportional Valves: Selection and Application*, Michael J. Tonyan, edited by Tobi Goldoftas
40. *Pressure Gauge Handbook*, AMETEK, U.S. Gauge Division, edited by Philip W. Harland
41. *Fabric Filtration for Combustion Sources: Fundamentals and Basic Technology*, R. P. Donovan
42. *Design of Mechanical Joints*, Alexander Blake
43. *CAD/CAM Dictionary*, Edward J. Preston, George W. Crawford, and Mark E. Coticchia
44. *Machinery Adhesives for Locking, Retaining, and Sealing*, Girard S. Haviland
45. *Couplings and Joints: Design, Selection, and Application*, Jon R. Mancuso
46. *Shaft Alignment Handbook*, John Piotrowski
47. *BASIC Programs for Steam Plant Engineers: Boilers, Combustion, Fluid Flow, and Heat Transfer*, V. Ganapathy
48. *Solving Mechanical Design Problems with Computer Graphics*, Jerome C. Lange
49. *Plastics Gearing: Selection and Application*, Clifford E. Adams
50. *Clutches and Brakes: Design and Selection*, William C. Orthwein
51. *Transducers in Mechanical and Electronic Design*, Harry L. Trietley
52. *Metallurgical Applications of Shock-Wave and High-Strain-Rate Phenomena*, edited by Lawrence E. Murr, Karl P. Staudhammer, and Marc A. Meyers
53. *Magnesium Products Design*, Robert S. Busk
54. *How to Integrate CAD/CAM Systems: Management and Technology*, William D. Engelke
55. *Cam Design and Manufacture: Second Edition*; with cam design software for the IBM PC and compatibles, disk included, Preben W. Jensen
56. *Solid-State AC Motor Controls: Selection and Application*, Sylvester Campbell
57. *Fundamentals of Robotics*, David D. Ardayfio
58. *Belt Selection and Application for Engineers*, edited by Wallace D. Erickson
59. *Developing Three-Dimensional CAD Software with the IBM PC*, C. Stan Wei
60. *Organizing Data for CIM Applications*, Charles S. Knox, with contributions by Thomas C. Boos, Ross S. Culverhouse, and Paul F. Muchnicki

61. *Computer-Aided Simulation in Railway Dynamics*, by Rao V. Dukkipati and Joseph R. Amyot
62. *Fiber-Reinforced Composites: Materials, Manufacturing, and Design*, P. K. Mallick
63. *Photoelectric Sensors and Controls: Selection and Application*, Scott M. Juds
64. *Finite Element Analysis with Personal Computers*, Edward R. Champion, Jr., and J. Michael Ensminger
65. *Ultrasonics: Fundamentals, Technology, Applications: Second Edition, Revised and Expanded*, Dale Ensminger
66. *Applied Finite Element Modeling: Practical Problem Solving for Engineers*, Jeffrey M. Steele
67. *Measurement and Instrumentation in Engineering: Principles and Basic Laboratory Experiments*, Francis S. Tse and Ivan E. Morse
68. *Centrifugal Pump Clinic: Second Edition, Revised and Expanded*, Igor J. Karassik
69. *Practical Stress Analysis in Engineering Design: Second Edition, Revised and Expanded*, Alexander Blake
70. *An Introduction to the Design and Behavior of Bolted Joints: Second Edition, Revised and Expanded*, John H. Bickford
71. *High Vacuum Technology: A Practical Guide*, Marsbed H. Hablanian
72. *Pressure Sensors: Selection and Application*, Duane Tandeske
73. *Zinc Handbook: Properties, Processing, and Use in Design*, Frank Porter
74. *Thermal Fatigue of Metals*, Andrzej Weronski and Tadeusz Hejwowski
75. *Classical and Modern Mechanisms for Engineers and Inventors*, Preben W. Jensen
76. *Handbook of Electronic Package Design*, edited by Michael Pecht
77. *Shock-Wave and High-Strain-Rate Phenomena in Materials*, edited by Marc A. Meyers, Lawrence E. Murr, and Karl P. Staudhammer
78. *Industrial Refrigeration: Principles, Design and Applications*, P. C. Koelet
79. *Applied Combustion*, Eugene L. Keating
80. *Engine Oils and Automotive Lubrication*, edited by Wilfried J. Bartz
81. *Mechanism Analysis: Simplified and Graphical Techniques, Second Edition, Revised and Expanded*, Lyndon O. Barton
82. *Fundamental Fluid Mechanics for the Practicing Engineer*, James W. Murdock
83. *Fiber-Reinforced Composites: Materials, Manufacturing, and Design, Second Edition, Revised and Expanded*, P. K. Mallick
84. *Numerical Methods for Engineering Applications*, Edward R. Champion, Jr.
85. *Turbomachinery: Basic Theory and Applications, Second Edition, Revised and Expanded*, Earl Logan, Jr.
86. *Vibrations of Shells and Plates: Second Edition, Revised and Expanded*, Werner Soedel
87. *Steam Plant Calculations Manual: Second Edition, Revised and Ex panded*, V. Ganapathy
88. *Industrial Noise Control: Fundamentals and Applications, Second Edition, Revised and Expanded*, Lewis H. Bell and Douglas H. Bell
89. *Finite Elements: Their Design and Performance*, Richard H. MacNeal
90. *Mechanical Properties of Polymers and Composites: Second Edition, Revised and Expanded*, Lawrence E. Nielsen and Robert F. Landel
91. *Mechanical Wear Prediction and Prevention*, Raymond G. Bayer

92. *Mechanical Power Transmission Components*, edited by David W. South and Jon R. Mancuso
93. *Handbook of Turbomachinery*, edited by Earl Logan, Jr.
94. *Engineering Documentation Control Practices and Procedures*, Ray E. Monahan
95. *Refractory Linings Thermomechanical Design and Applications*, Charles A. Schacht
96. *Geometric Dimensioning and Tolerancing: Applications and Techniques for Use in Design, Manufacturing, and Inspection*, James D. Meadows
97. *An Introduction to the Design and Behavior of Bolted Joints: Third Edition, Revised and Expanded*, John H. Bickford
98. *Shaft Alignment Handbook: Second Edition, Revised and Expanded*, John Piotrowski
99. *Computer-Aided Design of Polymer-Matrix Composite Structures*, edited by Suong Van Hoa
100. *Friction Science and Technology*, Peter J. Blau
101. *Introduction to Plastics and Composites: Mechanical Properties and Engineering Applications*, Edward Miller
102. *Practical Fracture Mechanics in Design*, Alexander Blake
103. *Pump Characteristics and Applications*, Michael W. Volk
104. *Optical Principles and Technology for Engineers*, James E. Stewart
105. *Optimizing the Shape of Mechanical Elements and Structures*, A. A. Seireg and Jorge Rodriguez
106. *Kinematics and Dynamics of Machinery*, Vladimír Stejskal and Michael Valášek
107. *Shaft Seals for Dynamic Applications*, Les Horve
108. *Reliability-Based Mechanical Design*, edited by Thomas A. Cruse
109. *Mechanical Fastening, Joining, and Assembly*, James A. Speck
110. *Turbomachinery Fluid Dynamics and Heat Transfer*, edited by Chunill Hah
111. *High-Vacuum Technology: A Practical Guide, Second Edition, Revised and Expanded*, Marsbed H. Hablanian
112. *Geometric Dimensioning and Tolerancing: Workbook and Answerbook*, James D. Meadows
113. *Handbook of Materials Selection for Engineering Applications*, edited by G. T. Murray
114. *Handbook of Thermoplastic Piping System Design*, Thomas Sixsmith and Reinhard Hanselka
115. *Practical Guide to Finite Elements: A Solid Mechanics Approach*, Steven M. Lepi
116. *Applied Computational Fluid Dynamics*, edited by Vijay K. Garg
117. *Fluid Sealing Technology*, Heinz K. Muller and Bernard S. Nau
118. *Friction and Lubrication in Mechanical Design*, A. A. Seireg
119. *Influence Functions and Matrices*, Yuri A. Melnikov
120. *Mechanical Analysis of Electronic Packaging Systems*, Stephen A. McKeown
121. *Couplings and Joints: Design, Selection, and Application, Second Edition, Revised and Expanded,* Jon R. Mancuso
122. *Thermodynamics: Processes and Applications*, Earl Logan, Jr.
123. *Gear Noise and Vibration*, J. Derek Smith
124. *Practical Fluid Mechanics for Engineering Applications,* John J. Bloomer
125. *Handbook of Hydraulic Fluid Technology*, edited by George E. Totten
126. *Heat Exchanger Design Handbook*, T. Kuppan

127. *Designing for Product Sound Quality*, Richard H. Lyon
128. *Probability Applications in Mechanical Design*, Franklin E. Fisher and Joy R. Fisher
129. *Nickel Alloys,* edited by Ulrich Heubner
130. *Rotating Machinery Vibration: Problem Analysis and Troubleshooting*, Maurice L. Adams, Jr.
131. *Formulas for Dynamic Analysis,* Ronald L. Huston and C. Q. Liu
132. *Handbook of Machinery Dynamics*, Lynn L. Faulkner and Earl Logan, Jr.
133. *Rapid Prototyping Technology: Selection and Application,* Kenneth G. Cooper
134. *Reciprocating Machinery Dynamics: Design and Analysis*, Abdulla S. Rangwala
135. *Maintenance Excellence: Optimizing Equipment Life-Cycle Decisions,* edited by John D. Campbell and Andrew K. S. Jardine
136. *Practical Guide to Industrial Boiler Systems*, Ralph L. Vandagriff
137. *Lubrication Fundamentals: Second Edition, Revised and Expanded*, D. M. Pirro and A. A. Wessol
138. *Mechanical Life Cycle Handbook: Good Environmental Design and Manufacturing,* edited by Mahendra S. Hundal
139. *Micromachining of Engineering Materials,* edited by Joseph McGeough
140. *Control Strategies for Dynamic Systems: Design and Implementation,* John H. Lumkes, Jr.
141. *Practical Guide to Pressure Vessel Manufacturing,* Sunil Pullarcot

Additional Volumes in Preparation

Diesel Engine Engineering: Dynamics, Design, and Control, Andrei Makartchouk

Reliability Verification, Testing, and Analysis in Engineering Design, Gary S. Wasserman

Nondestructive Evaluation, edited by Peter J. Shull and Michael Ehrlich

Bearing Design in Machinery, Avraham Harnoy

Handbook of Machine Tool Analysis, Ioan D. Marinescu, Constantin Ispas, and Dan Boboc

Implementing Concurrent Engineering in Small Companies, Susan Carlson Skalak

Mechanical Properties of Engineering Materials, Wole Soboyejo

Mechanical Engineering Software

Spring Design with an IBM PC, Al Dietrich

Mechanical Design Failure Analysis: With Failure Analysis System Software for the IBM PC, David G. Ullman

Practical Guide to Pressure Vessel Manufacturing

Sunil Pullarcot
FACT Engineering and Design Organisation
Kerala, India

MARCEL DEKKER, INC. NEW YORK • BASEL

ISBN: 0-8247-0740-0

This book is printed on acid-free paper.

Headquarters
Marcel Dekker, Inc.
270 Madison Avenue, New York, NY 10016
tel: 212-696-9000; fax: 212-685-4540

Eastern Hemisphere Distribution
Marcel Dekker AG
Hutgasse 4, Postfach 812, CH-4001 Basel, Switzerland
tel: 41-61-261-8482; fax: 41-61-261-8896

World Wide Web
http://www.dekker.com

The publisher offers discounts on this book when ordered in bulk quantities. For more information, write to Special Sales/Professional Marketing at the headquarters address above.

Current printing (last digit):
10 9 8 7 6 5 4 3 2 1

PRINTED IN THE UNITED STATES OF AMERICA

Preface

Twenty years ago, after completing my degree in mechanical engineering, I began my career in a pressure vessel fabrication company. Because I was a newcomer to the industry, I sought a reliable reference explaining fabrication and the corresponding procedures of quality assurance and control (QAC). I came across books such as *Pressure Vessel Design Manual*, by Dennis R. Moss, and *Pressure Vessel Handbook*, by Eugene F. Megyesy, but these were confined to aspects of design. The ASME code and the other national codes were the only documents available for reference.

The ASME code provided all the expected specifications but failed to suggest the means of achieving compliance with such specifications—perhaps because it was intended for engineers with an insight into the underlying logic of the code. Thus, my efforts to understand the code led only to further confusion.

This book summarizes my efforts, over 20 years, to bridge the gap between code specifications and the procedures necessary to comply with them. It can be considered a guide to and clarification of the ASME code—encompassing fabrication, inspection, testing, and documentation. Because the book offers wide coverage of fabrication (from organization setup to final testing and documentation), it also promotes uniformity in manufacturing, inspection, testing, and documentation.

Chapters 1 through 10, which are based on my experience in QAC and heavy fabrication, are dedicated to practical aspects of fabrication, beginning with organizational requirements and leading up to the final documentation of the vessel. Chapters 11 to 16 describe procedures that are essential for conducting various inspections and tests. Chapter 17 addresses welding procedure specifications, the procedure qualification record, and welder qualification tests. Chapters 18 and 19 detail a typical quality plan and an inspection and test plan, which is also included in QAC documentation. Chapter 20 provides recommended tolerances for vessels; such tolerances, which vary only slightly between most consultants, are usually not given in the code. Chapter 21 consists of excerpts from ASME Section VIII Div(1) (1998 edition), which is frequently referred to during manufacture. Because the code is continually updated to incorporate the latest available data, users of this book are encouraged to refer to these sections from the latest editions of the code as they become available.

The book is written based on the requirements of ASME Section VIII Div(1). Other sections of the ASME code, such as V and IX, are also referred to for the full coverage. Even though the book is based only on ASME codes, it could be extended to other codes for pressure vessels, because the relationship between fabrication and nondestructive testing (NDT) is the same regardless of the code used.

I would like to acknowledge the assistance and cooperation I have received from the learned surveyors of Lloyd's Register, ABS Industrial Verification, Det Norske Veritas, and Hartford Steam Boiler. The continued interaction I have had with these third-party inspection agencies has been a tremendous help in giving this book its present shape. Similarly, the technical and moral support extended to me by my colleagues at FACT Engineering and Design Organisation (FEDO) has played a vital role in my completion of this work.

Sunil Pullarcot

Contents

Practical Guide to Pressure Vessel Manufacturing

1

Introduction

1 GENERAL

Pressure vessels are manufactured all over the world under various international standards. These standards do specify the requirements to be met during the manufacture of pressure vessels; however, the means used to meet these requirements are not specified in any of these standards. In the following chapters, a detailed methodology to check each essential parameter specified in the relevant standards is established. Even though the attempt is made based on ASME Section VIII Div (1), the same methodology could also be extended to pressure vessels manufactured under different standards. The only difference would be in the acceptability norms that obviously could be gathered from the applicable standard.

2 PRESSURE VESSEL

A vessel used for the containment of pressure (either internal or external) is a pressure vessel. Pressure vessels can be of any shape. However, the most complicated one, as envisaged in ASME Section VIII Div (1), is shown in Figure 1.1.

3 STANDARDS

The most commonly used standard in the manufacture of pressure vessels in India is ASME Section VIII Div (1) even though there is another Indian standard for unfired pressure vessels. The standards that are commonly used in India for the manufacture of pressure vessels are the following.

1. ASME Section VIII Div (1): Rules for Construction of Unfired Pressure Vessels
2. ASME Section VIII Div (2): Alternative Rules for Construction of Unfired Pressure Vessels
3. ASME Section VIII Div (3): Alternative Rules for Construction of High Pressure Vessels
4. IS 2825: Indian Standard code for Unfired Pressure Vessels
5. BS 5500: Specification for Unfired Fusion Welded Pressure Vessels
6. Chapter J of Rules and Regulations for the Construction and Classification of Steel Ships by Lloyds Register of Shipping
7. AD Merkblater Technical Rules for Pressure Vessels

4 METHODOLOGY

The basic information for designing pressure vessels, being a part of the process plant, should be formulated by the process engineering group and contain such criteria as shown in Figure 1.2. This is known as the design data sheet (DDS).

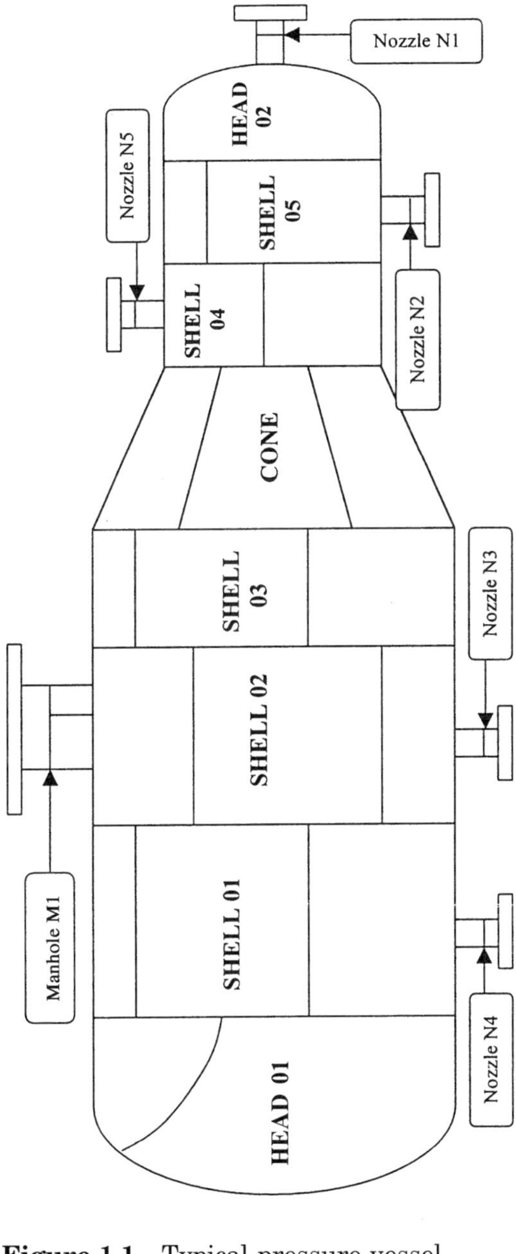

Figure 1.1 Typical pressure vessel.

1 Customer/Client				
2 Customer Order No.				
3 Shop Order No.				
4 Design Drawing				
5 Specifications				
6 Vessel Name				
7 Equipment No.				
8 Design Code & Addenda				
9 Design Pressure	Internal		External	
10 Design Temperature				
11 Operating Pressure	Internal		External	
12 Operating Temperature				
13 Vessel Diameter				
14 Volume				
15 Contents & Specific Gravity				
16 Service				
17 Design Liquid Level				
18 Maximum Allowable Working Pressure				
19 Test Pressure	Shop		Field	
20 Heat Treatment				
21 Joint Efficiency				
22 Corrosion Allowance Shell				
Heads				
Nozzles				
Boot				
23 Materials Shell				
Heads				
Nozzles				
Flanges				
Bolting				
Supports				
24 Flange Ratings				
25 Weights	Fabricated		Operating	
	Empty		Test	
26 Notes / Remarks				

Figure 1.2 Design data sheet.

Additional conditions that are to be considered during design, manufacture, and testing, along with technical/commercial conditions of delivery, will be furnished in a document called the technical procurement specifications (TPS), which is part of the purchase order (PO).

Based on the design data sheet and other technical conditions, the design department of the fabricator performs the design calculations as per the applicable code, and a detailed fabrication drawing is prepared. This in turn is sent to the client, consultant, third party inspection agency, statutory authority, etc., for their scrutiny and approval.

After the approval of the design and drawings by various agencies (primarily the client or consultant), the planning department of the fabricator generates material requisitions to the procurement department. The material requisition is just a translation of all the relevant data given in the drawing as well as in the TPS in a form that can be understood by the vendor for that particular item.

The purchase department floats enquiry for materials to reputed vendors who are already registered with the fabricator. The department is primarily responsible for obtaining materials of required quality at the lowest price and within the stipulated time frame. The responsibility of the department ceases only when the material is received at stores and accepted by the quality assurance and control (QAC) department.

On receipt of the material, the planning department issues a work order to the production department with authorization to draw materials from stores along with certified-for-fabrication drawings and a detailed delivery schedule for the work.

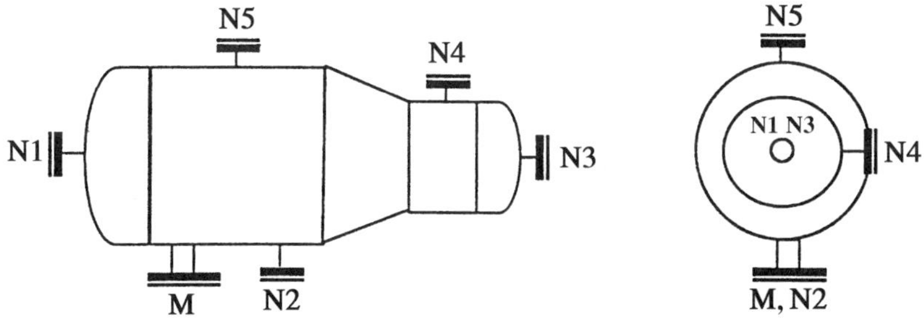

Figure 1.3 Sketch of two views of pressure vessel.

The production department in turn works out its microlevel planning based on availability of manpower and machines to meet the deadlines committed to the client.

The microlevel planning of the QAC department is carried out in such a way that it shall not conflict with the overall production schedule nor the microplanning of production.

With the concerted efforts of all the mentioned departments, as well as the other service departments like stores, maintenance, personnel, finance, etc., the targets for cost, time, and quality can be met. If these targets are achieved, it paves way for prosperity of the organization.

2

Organization

1 THE NEED AND ROLE OF ORGANIZATION

In order to achieve the required level of quality in production, all the departments mentioned in Chapter 1 must function in a concerted and organized manner with the ultimate aim of better and consistent overall quality for the product. This in turn calls for a strong and well-defined organization well interlaced between departments so that all departments function as a team with a single corporate objective, rather than focus on the best performance of any individual department. The product quality shall in no case be lower than that desired by the client in the specifications. At the same time it need not be of extremely high quality, in which case the product cost would also be very high. The organization shall be capable of delivering the products of optimal quality at the lowest cost. Therefore, repairs and reworks shall be avoided to the extent possible as they inevitably turn out to be very expensive in terms of both money and time. With this objective in mind, this chapter describes a successful organizational model based on the author's experience in the field.

2 ORGANIZATION CHART

A typical organization chart for a medium scale pressure vessel manufacturing setup is shown in Figure 2.1.

The positions shown in the organization chart can be broadly classified into three levels. The level consisting of the General Manager (GM) and the Deputy General Managers (DGMs) is called the top management; they make the policy and strategic decisions. The middle management level consists of Managers and Assistant Managers, who make decisions on day-to-day affairs. They are also expected to brief the top management with an abridged version of all decisions and the results of same for information management to be utilized for future policy decisions. Interdepartmental coordination is another major area of concern for middle managers, and they must communicate the goals set by top management to facillitate implementation by the junior managers of the company. The junior managers include engineers and supervisors. The engineers mostly carry out the theoretical and analytical part of the work, whereas the supervisors are expected to provide the labor management.

Out of the positions described, engineers shall man all engineering and procurement disciplines, whereas the officers and staff in the finance, personnel, and stores departments may be from the nontechnical side.

Depending on the turnover of the company and amount of work, additional posts can be generated as needed. In the

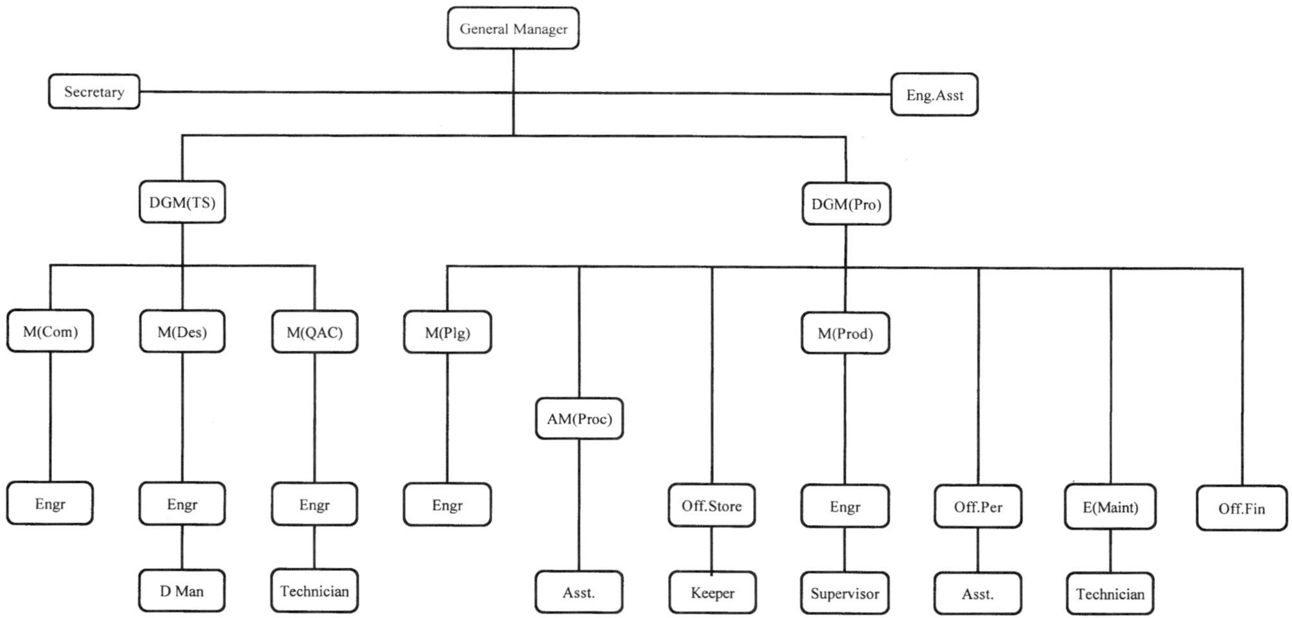

1) All levels below level III shall be workmen in respective desciplines.
2) As the nature of job is not continuous, posts below level III may be augmented on hire depending on need
3) Legends:-

DGM(TS)	Dy.General Manager(Technical Services)	M(QAC)	Manager(Quality Assurance & Control)	E(Maint)	Engineer(Maintenance)
DGM(Pro)	Dy.General Manager(Production)	M(Plg)	Manager(Planning)	Eng Asst	Engineering assistant to GM
M(Com)	Manager(Commercial)	M(Prod)	Manager(Production)	Off(Stores)	Officer(Stores)
M(Des)	Manager(Design & Drawing)	AM(Proc)	Asst. manager(Procurement)	Off(Per)	Officer(Personnel)
Off(Fin)	Officer(Finance)				

Figure 2.1 Organization chart.

interest of the organization, it is better to have all key personnel from every department on a permanent basis, and the junior staff required can be hired on a contract basis depending on need.

3 DEPARTMENT FUNCTIONS AND RESPONSIBILITIES

3.1 Commercial Department

The commercial department is responsible for bagging orders for the firm by constantly keeping in touch with probable and potential clients. Its scope of work is vast and a few of the salient responsibilities are as follows:

1. Keep the company in the approved vendor list of all prospective clients as well as renowned consultants.
2. Obtain tender documents and make a competitive offer, attending to technical discussions and clarifications along with concerned engineers from the required departments; follow up on offers; and bag orders. (In order to prepare estimates, preliminary designing may be done by the design department and may be discussed with other departments to confirm on all practicalities of the job in formal pretender meetings.)
3. Enter into formal contracts with clients and consultants as needed to obtain orders.
4. Transfer purchase order (PO) and other documents pertaining to the contract to the concerned departments and explain to them the basis of estimation in a postorder meeting.
5. Keep track of changes in the environment and direct company in equipping itself for meeting new challenges as well as to keep the company always in the forefront.

6. Promote the company in all appropriate potential avenues.
7. Prepare and monitor the schedules for improvements and developmental activities as well as diversification plans.

3.2 Design and Drawing Department

The first step in the manufacture of pressure vessel is the design of the vessel based on the design data sheet as well as the other technical conditions stipulated in the technical procurement specifications (TPS). For this reason, the department must go through all the technical conditions while carrying out the design. As all the subsequent activities are based on this fabrication drawing, all practical aspects of production must be considered. Some of the major responsibilities of the department are

1. Carry out the design as per applicable codes and client specifications; prepare fabrication drawings; submit design calculations and fabrication drawings for the approval of the client, third party, consultant, and statutory authority to obtain their approvals as needed.
2. Issue the prepared drawings to all the concerned departments to obtain feedback as to the method of fabrication and sequence of fabrication and incorporate such changes in the drawing as desired by the departments and obtain approvals for the changes effected.
3. Issue the certified-for-fabrication drawings to various departments to start the work.
4. Maintain reference standards like codes, etc., for the design and fabrication and keep abreast with the latest design and drawing trends.
5. Obtain as-built dimensions from concerned departments and prepare as-built drawings.
6. Issue solutions to fabrication problems and for mate-

rial problems and suggest alternative materials and methods when needed.

3.3 Planning Department

This department is the backbone of the production even though they are not directly involved in the production function. A few major responsibilities of the department are as follows:

1. Keep track of all materials in hand and plan the most effective and economical use of same.
2. Generate material requisitions to procurement department for raw materials other than that available in the stores, highlighting the required delivery date. The format suggested for this purpose is shown in Figure 2.2.
3. Issue technical recommendations on offers obtained by the procurement department and keep track of same up to their delivery to stores.
4. Issue work order and material authorization for shops to draw materials and to start with the actual execution of the job. The suggested formats are shown in Figures 2.3 and 2.4.
5. Schedule the shop loading based on machine and manpower availability considering all other aspects of economy.
6. Identify and monitor critical path activities and keep management posted on schedules, targets, and tasks.
7. Schedule necessary discussions with clients including progress reporting.
8. Coordinate with other departments in overcoming problems.
9. Conduct periodic review meetings with concerned departments and chart out appropriate action plans as needed.

EMBLEM

MATERIAL REQUISITION

Order No.		M R No.	
Purchaser		Date	
Description		Delivery date	

The following items may be procurred for the above referred project as per the delivery schedule.

Sl.No.	Description	Applicable Specification	Quantity	Unit	Due Date

(Note: - As far as possible, one request may be used for one type of material only)

Manager Planning

Routing: -

Asst. Manager Procurement

Cc: Manager (Production)
Manager (QAC)
O/C 1: Master File
O/C 2: Job File

Figure 2.2 Format for material requisition.

EMBLEM

WORK ORDER

Order No.				Work Order No.	
Purchaser				Date	
Description				Quantity	
Drawing Numbers					

The production is hereby requested to manufacture ----number(s) of pressure vessels/ columns/ as per the above drawings with the salient features as below.

1	Vessel Identification Number(s)	
2	Design & Manufacturing Code	
3	Approximate Diameter & Length	
4	Design Pressure & Temperature	
5	Test Pressure	
6	Heat Treatment	
7	Joint Efficiency	
8	NDT requirements	
9	Material of Construction	
10	Inspection Agency	
11	Cleaning & Painting	
12	As Built Documents	

Manager Planning

Routing: -

Manager Production

Cc: Manager (QAC)
Officer (Stores)
O/C 1: Master File
O/C 2: Job File

Figure 2.3 Format for work order.

EMBLEM

MATERIAL AUTHORIZATION

Order No.		Authorization No.	
Purchaser		Date	
Description		Work Order Ref:	

The production is hereby authorized to draw the following materials received at stores as per the references indicated against each for the above referred work order. The scrap material against this project should be returned to stores on completion of the work with material reconciliation statement.

Sl.No.	Description	Quantity	Unit	PO & RR Details

Manager Planning

Routing: -

Manager Production

Cc: Manager (QAC)
Officer (Stores)
O/C 1: Master File
O/C 2: Job File

Figure 2.4 Format for material authorization.

3.4 Procurement Department

The basic function of this department is to procure materials as specified in the requisition of the planning department, within the allocated budget and time. A few major functions of the department are

1. Maintain an up-to-date approved vendor list on all raw materials, consumables, and spares for the heavy machinery in the shops.
2. Obtain enquiries within the time frame specified by the planning department as per the technical procurement specifications and at the lowest cost.
3. Place orders with reliable parties and expedite them to ensure timely receipt at stores.
4. Collect the documents pertaining to supply and forward them to the concerned departments for further action.
5. Constantly upgrade and revise vendors list based on previous interactions with these parties as well as on reliable information from various sources.

3.5 Stores

All raw materials received and accepted will be under the custody of stores. They are expected to issue the same to production based on authorization from the planning department. Listed here are a few of their salient functions:

1. Receive and store materials at the incoming stores area earmarked for the purpose.
2. Offer materials to the quality assurance and control department (QAC) for the inspection of same through intimation of material receipt (a sample format is given in Figure 2.5).
3. Take all accepted materials to the stock and store them properly. The intimation of the materials shall

EMBLEM

INTIMATION OF MATERIAL RECIEPT

Purchase Order No.		Date	
Vendor		Date of Receipt	

DETAILS OF ITEMS RECEIVED

Sl.No.	Tag No.	Description

Location:
Store Keeper in Charge:
Status of MTC: Received/ not received

Officer (Stores):

Routing:-
Manager (QC&A)

Cc:	Manager (Purchase)	O/C 1	Master File
	Manager (Planning)	O/C 2	PO File
	Manager (Production)		

Figure 2.5 Intimation of material receipt.

be given to the planning department for further action.
4. Issue materials to users based on authorization from the planning department (for raw materials only).

3.6 Quality Assurance and Control Department

Even though the quality assurance and control department is a service department to production, it has a higher responsibility to the client by virtue of its function. Therefore the department enjoys more autonomy than others and shall be manned by well-qualified personnel with a high degree of integrity. Listed here are some of their important responsibilities:

1. Carry out the inspection of all incoming raw materials at stores based on PO terms and applicable codes as specified in the PO and maintain records to that effect.
2. Prepare and obtain approval for all quality assurance/quality control procedures, including the quality assurance plan (QAP), inspection and test plan (ITP), welding procedure specifications (WPS), procedure qualification records (PQRs), and the welder's qualification test (WQT), as well as nondestructive testing (NDT), postweld heat treatment (PWHT), and final hydrostatic testing (HT) procedures. All are based on requirements of applicable codes and contracted specifications.
3. Carry out stagewise inspection during manufacture and maintain records.
4. Carry out NDT and DT as required at various stages, reporting on unacceptable defects; review rectification of defects previously detected through rigorous follow-ups with production; and maintain records to this effect.

5. Prepare NDT summary, weld summary, etc., along with weld maps and consolidation of same to check the completeness of NDT.
6. Witness various tests carried out during the process of manufacture and prepare documents.
7. Coordinate with third party/client/consultant for witnessing the hold points/stages as laid out in the QAP and prepare documentation of same.
8. Issue manufacturer's certificates for the vessels in the required format.
9. Prepare manufacturer's data report (MDR) and submit it to client after approval by third party/consultant/client as required.
10. Supply information to the design department for the preparation of as-built drawings.
11. Perform quality auditing and reporting to top management on overall quality level of various departments directly involved in the making of the vessel, namely, design and drawing, planning, procurement, stores, production, and QAC departments.

3.7 Production Department

The physical transformation of raw materials to a pressure vessel is the function of the department, even though a lot of effort by other departments also goes into it prior to the start of manufacture. Listed here are some of the main functions of the production section:

1. Production of shells, dished ends, cones, subassemblies as per engineering drawings, code practices, and the schedule of the planning department.
2. Feed back to planning on daily production status, problems, remedial measures, catch-up plans, etc.
3. Offer materials or products to QAC at various stages of fabrication for inspection.

4. Return balance materials to stores with proper identification and certification by the inspection department or agency concerned.
5. Perform the physical conduct of tests other than NDT/DT as per approved procedures and the instructions of QAC.
6. Keep up valid calibration certificates for all the measuring instruments.
7. Use proper procedures for the manufacture of components based on the process to be used.
8. Perform PWHT as per the approved procedure and in consultation with QAC on placement of thermocouples and on alterations to the cycle.
9. Perform the physical conduct of the hydrostatic test and the post–hydro test cleaning of the vessel.
10. Clean and paint the vessel as per the PO.
11. Pack the equipment for shipment to site as per PO.

3.8 Maintenance Department

Improving the availability of machines required for production is the main function of this department. A few other responsibilities as noted are also usually entrusted to them:

1. Keep all the machines in good order using proper preventive maintenance and scheduled shutdowns without affecting production.
2. Prepare material requisition for spares and consumables for the smooth operation of all the equipment including welding machines, grinders, etc., apart from heavy equipment.
3. Maintain all company vehicles for the transport of goods for production.

3.9 Personnel Department

All personnel matters are to be tackled by this department, including the following:

1. Recruitment, training, maintenance of personal records, renewal of contracts, sanction of increments, fixation of pay, etc.
2. Welfare measures for employees.
3. Office administration matters including housekeeping, postage, and other office facilities management.

3.10 Finance Department

Anything related to finance should be in the scope of the finance department's work, including

1. Resources planning and allocation of funds for different jobs based on profitability.
2. Accounting and its regularization and ratification with all statutory agencies.
3. Payroll for all employees.
4. Payment to all outside agencies on accounts of purchase of materials and services.

4 MASTER DOCUMENT FLOW DIAGRAM

The flow of documents between departments is shown in Figure 2.6, which depicts only the salient documents.

5 INPUTS AND OUTPUTS OF DEPARTMENTS

The important inputs and outputs of the departments described in Section 3 are summarized in Table 1. Inputs and outputs are categorized in two main groups—namely, one to outside agencies and one to other departments within the organization.

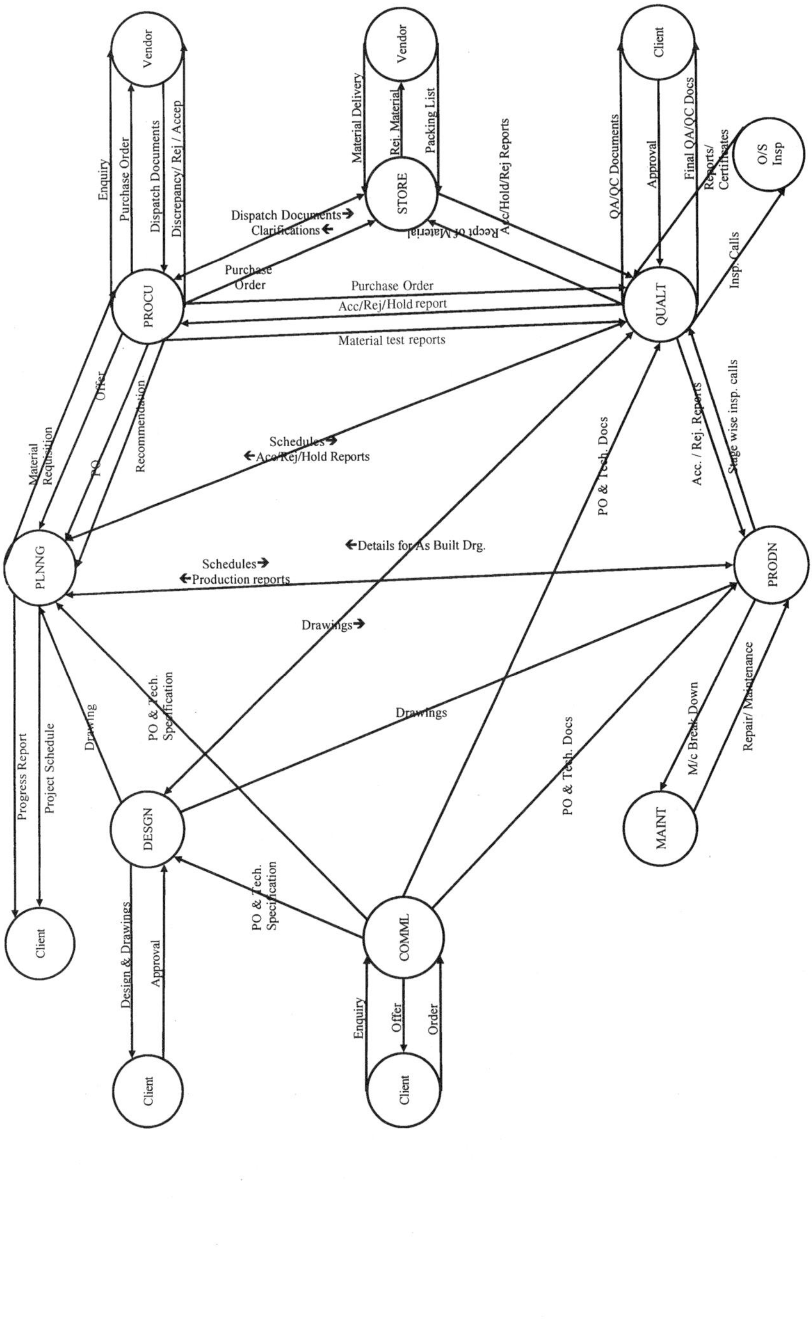

Figure 2.6 Master document flow diagram.

Table 1 Inputs and Outputs

Inputs		Outputs	
Outside the organization	Inside the organization	Outside the organization	Inside the organization
Commercial Department			
Enquiry	Preliminary design	Registration with clients	Intimation of order receipt
Order	Details for estimation	Offer	PO and technical specifications
Comments on schedules and reports	Suggestion for development activities	Clarification and negotiation	Schedules for development work
	Budgetary prices for cost estimation	Promotion	
	Fabrication details for cost estimation		
	Minutes of review meetings		
Design and Drawing Department			
Approval for design and drawing	NIT for preliminary design	Design and drawings for approval	Preliminary design for estimation
	PO and technical specifications	Final as-built drawings	Design and drawings for fabrication
	Problems during fabrication		Solutions to fabrication problems
	Dimensions for as-built drawings		
	Minutes of review meetings		
Planning Department			
Comments on schedules and reports	NIT and other documents	Schedules and progress reports	Fabrication details for cost estimation
Call for review meetings	PO and technical specifications	Minutes of review meetings	Schedules and progress reports
Material receipts and documents	Design and drawings for fabrication	Dispatch documents for completed vessel	Material requisition for raw materials

Table 1 (Continued)

Inputs		Outputs	
Outside the organization	Inside the organization	Outside the organization	Inside the organization
Planning Department			
	Credentials of vendors for evaluation	Invoices against the delivery	Technical recommendation on offers
	Materials status including new receipts		Materials request authorization
	Progress reports on materials and work		Periodic review of progress and plans
	Manpower and machine availability		Dispatch documents for completed vessel
	Inspection report of raw materials		Recommendation of vendors for inclusion
	Dispatch documents for completed vessel		
Procurement Department			
Credentials of vendors for evaluation	Recommendation of vendors for inclusion	Notice for vendor qualification	Credentials of vendors for evaluation
Offers from vendors for raw materials	Material requisition for raw materials	Intimation of qualification	Offers with comparison statement
Clarifications on enquiries	Technical recommendation on offers	Enquiry to vendor for raw materials	Dispatch documents and certificates
Dispatch documents and certificates	Acceptance or rejection of items	Clarifications on enquiries	PO and other conditions of delivery
Receipt of materials at stores	Request for purchase of spares	PO and other conditions of delivery	Dispatch documents and certificates
	Minutes of review meetings	Acceptance or rejection of items	

Inputs		Outputs	
Outside the organization	Inside the organization	Outside the organization	Inside the organization
Stores			
Receipt of materials at stores	PO and other conditions of delivery	Return of rejected materials	Inspection call to QAC
Dispatch documents and certificates	Dispatch documents and certificates		Issue of materials as per authorization
	Material inspection reports		
	Material request authorization		
	Minutes of review meetings		
Quality Assurance and Control Department			
Approved QAP, ITP, WPS, PQR, and WQT records	PO and technical specifications	QAP, ITP, WPS, PQR, and WQT records	WPS/PQR results
Approved NDT, PWHT, and HT procedures	Design and drawings for fabrication	NDT, PWHT, and HT procedures	PQR/WQT requirements
Certificate of inspection from third party (TP)	Schedules and progress reports	Stagewise inspection call to TP	Approved QAP, ITP, WPS, PQR, and WQT records
	Materials inspection call from stores	MDR with as-built drawing	Approved NDT, PWHT, and HT procedures
	Stagewise inspection call from shop		Materials inspection reports
	NDT requisition from shops as required		Stagewise inspection reports
	PWHT charts and reports		NDT reports with interpretation
	Dimensions for as-built drawings		Pad air test reports
	Minutes of review meetings		Endorsed PWHT charts and reports

Table 1 (Continued)

Inputs		Outputs	
Outside the organization	Inside the organization	Outside the organization	Inside the organization
Quality Assurance and Control Department			
			Hydraulic test reports signed by TP
			Dimensions for as-built drawings
			Cleaning and painting reports
			MDR with as-built drawing
			Quality audit report and analysis
Production Department			
	PO and technical specifications		Material request authorization
	Design and drawings for fabrication		Stagewise inspection call to QAC
	Schedules and progress reports		NDT requisition as required in code
	Material request authorization		Request to witness pad air test
	Advance notice on maintenance schedule		Request for PWHT verification
	Restrictions on machines		Request to witness hydrostatic test
	Minutes of review meetings		Dimensions for as-built drawings
			Dispatch documents for completed vessel
Maintenance Department			
Receipt of spares	Schedule of machine loading	Actions for maintenance contract	Advance notice on maintenance schedule

Inputs		Outputs	
Outside the organization	Inside the organization	Outside the organization	Inside the organization
Maintenance Department			
Maintenance contracts for special equipment	Intimation of breakdowns		Restrictions on machines
			Request for purchase of spares Consumables for machines
Personnel Department			
Liaison with statutory authorities	Manpower requirement	Liaison with statutory authorities	Renewal of contracts as required
	Training need for skilled workers		Granting of increments and promotions
	Employee records (personal and technical)		
Finance Department			
Audited financial results	Cashflow requirements Invoices and receipts for payment	Audited financial results	Salary statement of employees Payment to purchase of materials and services

3

Manufacture of Components

1 GENERAL

Pressure vessels consist of various components (as shown in Figure 3.1); based on the shape, various methodologies are adopted for manufacture. These methods must be strictly in accordance with the provisions of the prevailing code. The methods described here are in line with the requirements of ASME Section VIII Div (1) and hence shall meet almost all the requirements of other codes for pressure vessel manufacture.

2 COMPONENTS

The different components that are normally encountered in the manufacture of a pressure vessel are

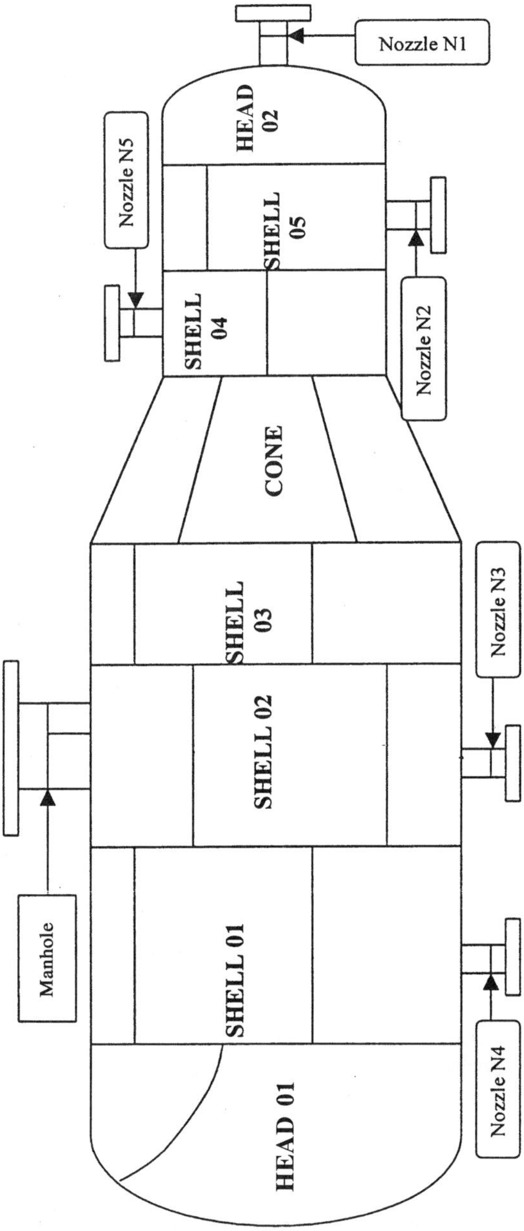

Figure 3.1 Typical pressure vessel.

1. Heads or dished ends
2. Shell (body of the vessel)
3. Cones or reducers
4. Attachments like nozzles, manholes, saddle supports, skirt supports, leg supports, trunnions, lifting lugs, platform and ladder supports, etc.

3 PRIORITY IN MANUFACTURE

The manufacture of heads and cones is considered to be more difficult than that of shells due to the difficulty in controlling the dimensions of these two items. (Whereas the dimensions of shell sections, for example, can be controlled precisely.) For this reason, if the vessel has dished ends or cones, they should be made first (of course, within the code tolerances) and shells are made later to suit these heads and cones.

4 MANUFACTURING PROCESSES

The basic manufacturing process adopted for the fabrication of the vessel is forming. Forming is the process by which the size or shape of the part is changed by application of force that produces stresses in the part which are greater than the yield strength and less than the fracture strength.

Depending on the temperature during fabrication, it is categorized as hot, warm, or cold forming. It is called hot working when the temperature is above the recrystallization temperature of the material, warm working if the temperature is sufficiently above room temperature but below recrystallization temperature, and cold working when the temperature is very much below recrystallization temperature, for example, at room temperature.

The manufacturing processes that fall under the forming operation specifically used in pressure vessel manufacturing are pressing, spinning, and bending.

4.1 Pressing

Pressing is the process usually adopted when the size of the part to be made is comparatively small. Domes and pipe caps are usually made using this process mostly at room temperature. In this process, the plate is gradually pressed to the required shape by local pressing, covering the entire area of the blank progressively. The thickness of the plate, the radius of curvature to be made, the material used, the machine capacity, and the availability of the right die for the shape are the limiting factors to decide on the process of manufacture. As formation is effected in stages, it is comparatively a slow process, and the shapes thus produced can have local deformations and hence the process calls for thorough inspection. In the case of cold pressing, due to the effect of work hardening, the chances for surface cracks on the outer surface is very high and hence calls for liquid penetrant testing (LPT) during examination of the surface of the component. It may even be necessary to carry out a heat treatment in between the cold forming in cases where the effects due to work hardening are severe. This is decided based on the calculated extreme fiber elongation expected in the component when it is cold formed. In hot pressing, utmost care shall be taken to see that the heating, the maintenance of heat during the process, and the subsequent cooling shall be strictly as per the procedure, as they have great bearing on the grain structure of the material. The temperature, heating/cooling rate, etc., are decided based on the material specification of the component as well as the amount of forming required, as specified in the applicable code.

4.2 Spinning

This process is used for making a variety of dished ends, namely, torispherical, 2:1 ellipsoidal, toriconical, and hemispherical. The specific advantages with this process are that the dished ends are made without the use of dies and it is the fastest process for the manufacture of dished ends.

Initially a hole of about 20 mm diameter is made, preferably by drilling, at the center of the blank, and the crown portion of the dish is formed using local pressing. Later it is loaded onto the spinning machine; then while spinning the partially formed blank, using adjustable guide rollers, the final shape is given. The movement of the guide roller shall be from the crown area progressing toward the straight flange through the knuckle area. As the blank is continuously spinning, the surface of the dished end shall be free from local deformations. However, there could be spinning lines all along the formed surface, especially at the knuckle portion, being the area under maximum deformation.

In this case also, based on the thickness of the blank, machine capacity etc., the process is carried out either cold or hot. In case of hot spinning, precautions are to be taken as for hot pressing mentioned in Section 4.1.

4.3 Bending

This process is conventionally called rolling even though it is not rolling. Rolling is a process where a thickness reduction to a plate takes place, whereas in bending only a curvature is given to the plate and no thickness reduction is expected.

The usual practice of bending is by passing the plate through either a three- or a five-roll plate bending machine of adequate capacity to bend the plate to the required diameter. The most commonly used bending machine in fabrication industry is the three-roll plate bending machine. A schematic diagram of a three-roll plate bending machine is shown in Figure 3.2. Prior to passing the complete length of the plate through the rollers, both edges of the plate are pressed to the required radius. This is called prepinging and is achieved by pressing the two power/drive rollers against the top idle roller and thereby bending the plate edge.

Once both the edges are pressed to the correct radius, the complete length of the plate is passed through the rollers to impart curvature in steps. If the diameter is large and the

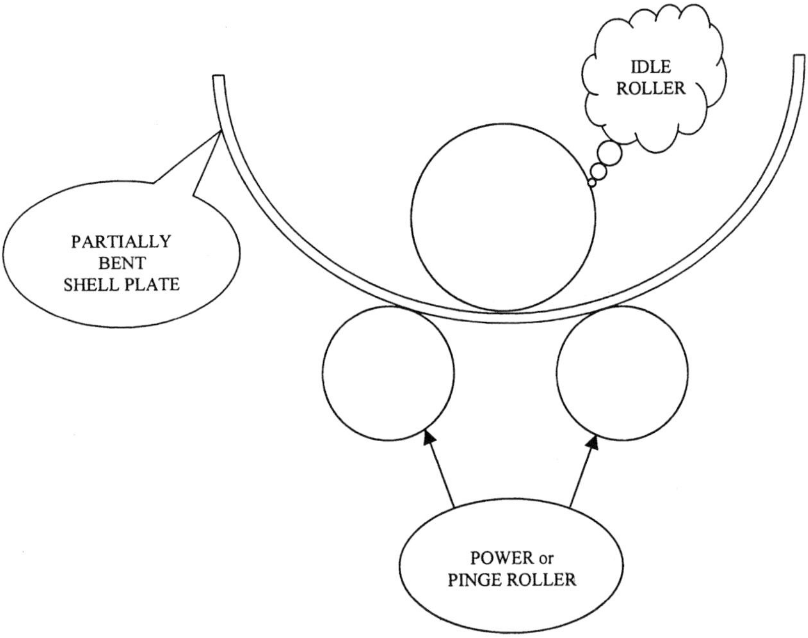

Figure 3.2 Schematic diagram of plate bending.

thickness is comparatively less (say up to 10 mm), then the curvature can be given in one pass. Whereas if the thickness is high and the diameter is small, the full bending is to be carried out in stages.

The profile of the plate is checked after bending using a template of required dimensions as highlighted in subsequent chapters of this book.

5 MANUFACTURE

5.1 General

The manufacture referred to here pertains to individual components that form a pressure vessel. Even though this is dealt with in detail in this chapter, a few aspects are repeated again in Chapter 4 for the sake of completeness and in order to better appreciate the importance of the sequence of manufacture.

5.2 Material Identification

For the manufacture of any component, the first step is to identify the material to be used for the job. For this the material test certificate of the intended material is essential. As all the raw materials have undergone incoming inspection at stores, it is presumed that detailed scrutiny was done at that point in time. However, verification as to the material specification and heat numbers stamped on the plate has to be carried out with respect to that given in the certificate and the specification in the drawing. If a few pressure parts are to be marked on the plate, say, dished end, cone, etc., then the identification on the plate has to be transferred to all the small components as shown in the diagram. Sufficiently large scrap plates shall also be identified in the same way, along with the personal stamp of the inspector responsible for the work. A certificate to this effect also shall be made, preferably with sketches for future reference. The form to be used for this is shown in Figure 3.3; Figure 3.4 indicates the actual work.

The mentioned identification shall be carried out for each and every pressure part of the vessel and, as far as possible, the material identification stamping shall be on the outside of the vessel so that they can be verified at any point of time in case of uninsulated vessels. The stamping on the vessel shall be carried out using low stress stamps. Hard punching shall be avoided.

When stamping is done on components like dished ends, care shall be taken to see that the stamping is made on the crown portion where the deformation is minimal. Likewise, in the case of subassemblies that are subjected to cold work after identification, the stamping shall be on an area which is subject to minimal cold work.

5.3 Heads

5.3.1 Flat Ends

Flat ends are usually made by forging. In that case it is a bought out item and is to be subjected to raw material inspection at the receiving end.

EMBLEM

CERTIFICATE OF MATERIAL IDENTIFICATION AND TRANSFER OF IDENTIFICATION

Order No.		Sheet	of
Purchaser		Attachment to Certificate No	
Description		Date	
Drawing No			

Part No(s). Marked												

Original Identification Found	Identification Transferred with Drawing No & Part No

SKETCH OF PLATE

Number of scrap pieces identified (with approx. size & shape) :

Material Marked by :

Material Identified by :

Figure 3.3 Certificate of material identification and transfer of identification.

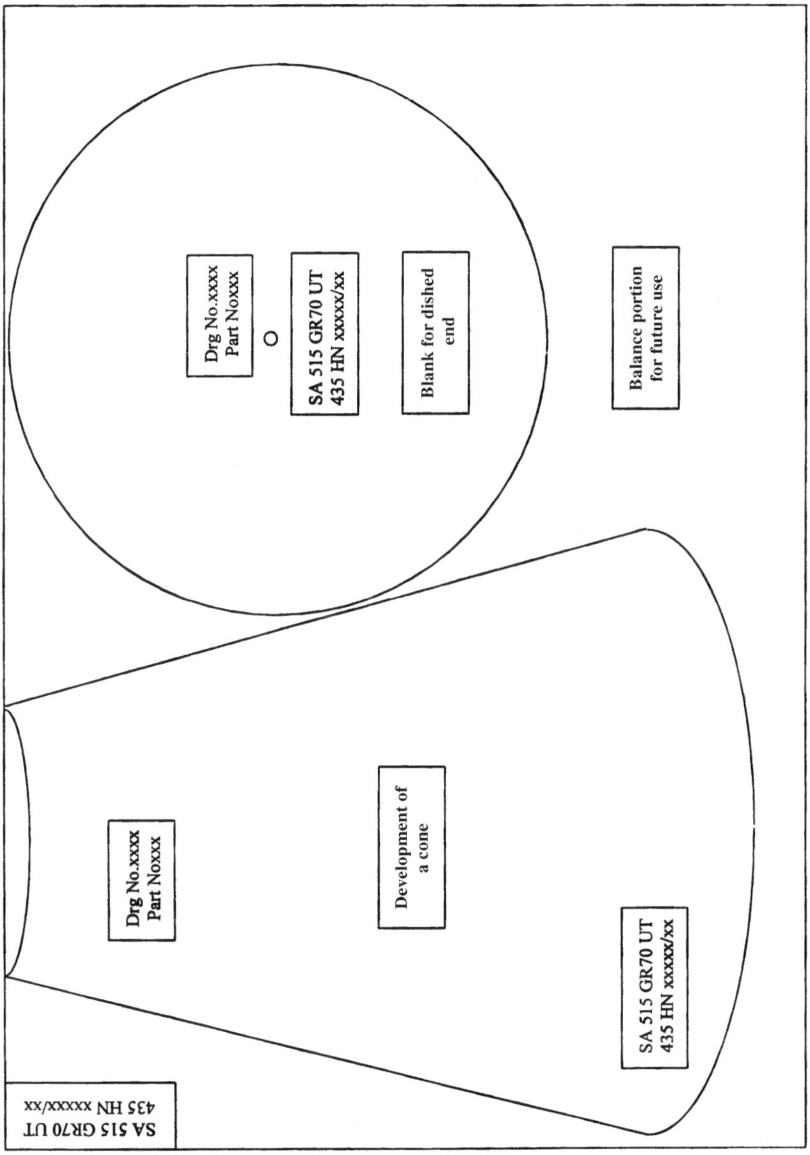

Figure 3.4 Plate marking and transfer of identification.

Semi-Ellipsoidal Dished Ends. This is the most commonly used type of dished end in the pressure vessel industry. The inside depth of the dished end excluding the height of the straight face is one-half the inside radius of the dish and hence is called 2:1 ellipsoidal.

Dished ends are manufactured by cold or hot pressing and in a combination process of pressing and spinning. Dished ends of comparatively smaller size and lower thickness, diameter up to around 700 mm with thickness over 10 mm, are made by either cold pressing or hot pressing, whereas all larger size dished ends are made by the combination process of both pressing and subsequent spinning.

Hemispherical Dished Ends. Dished ends of this category are also produced in the same way as that of 2:1 ellipsoidal heads.

It has already been mentioned that for spinning the dished end, a hole of approximately 20 to 30 mm in diameter has to be provided at the center of the blank for holding during spinning. In case any nozzles are to be attached at the center of the dished end, the opening is enlarged to fit the nozzle. If not, the hole has to be plugged, either by welding alone or by welding in a plug whose thickness will match the thickness of the dish at that place. In both cases, the hole is enlarged slightly to a maximum diameter of 60 mm as specified in UW 34 (see Chapter 21, Section 1 of this volume) and then plugged using matching material with the V groove as shown in Figure 3.5. A single V toward the inside of the dish is preferred when the thickness is below 16 mm. For larger thickness, an unequal V of $^2/_3\,t$ and $^1/_3\,t$ with the major V from inside is best suited from the distortion point of view.

Torispherical Dished Ends. This shape is also similar to that of 2:1 ellipsoidal. However, there is a difference in consideration at the design stage based on the L/r ratio which is not in the purview of this write up. As far as the manufacture is concerned, the methodology is one and the same as that of either hemispherical or 2:1 ellipsoidal heads.

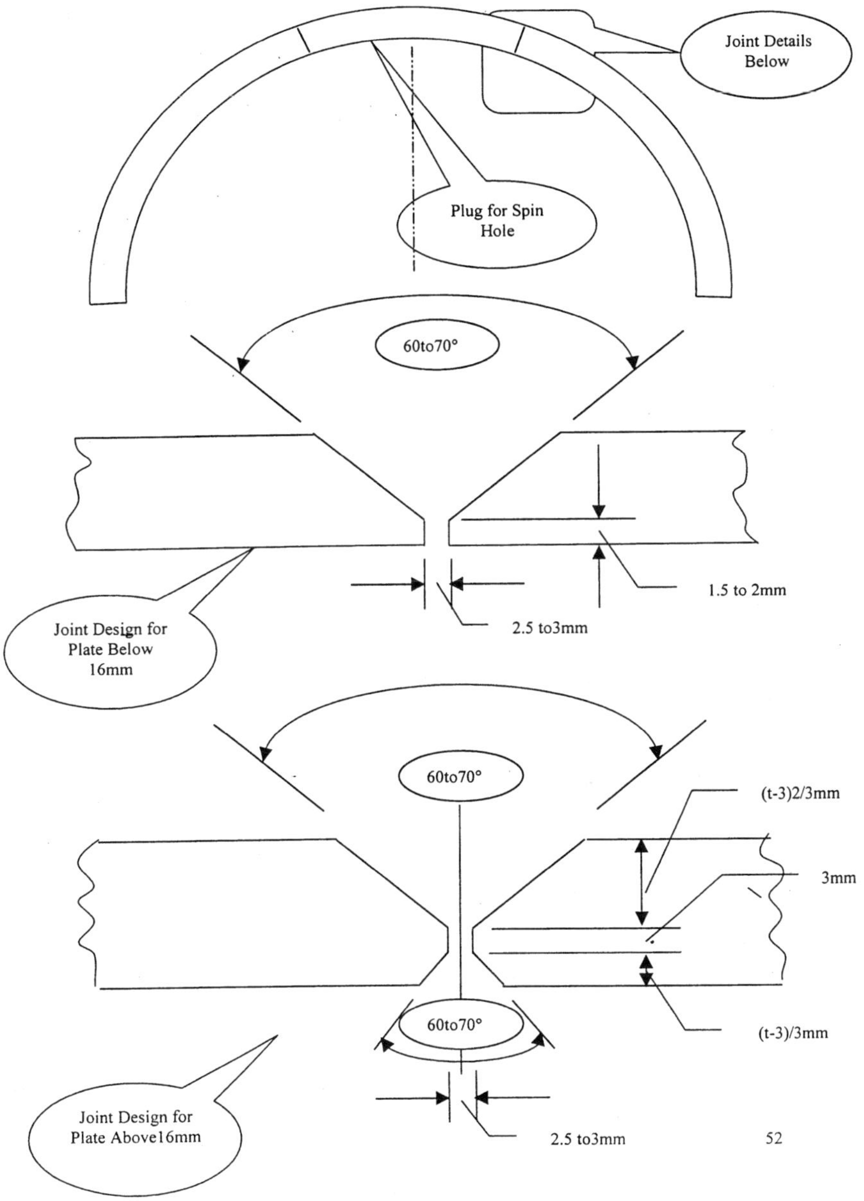

Figure 3.5 Spin hole plugging details.

5.4 Shells

5.4.1 Shells Made from Single Plate

If the length of the plate is sufficient to accommodate the full circumference of the shell, this is the preferred condition because the shell will have only one longitudinal seam. The plate is first cut to the required length and breadth and later on all four sides; a bevel is prepared as in the drawing. The square cutting and the bevel preparation is carried out by oxyacetylene flame for carbon steel plates and by plasma-arc process for stainless and high alloy steels. The cut edges are then ground back to sound metal at least by 1 to 1.5 mm so that all the adversely affected material in the cut zone is removed. This will help in preventing the oxidized metal from going into the weld pool, thereby making the weld incompatible to the base metal. The dressed edges are then examined for defects like laminations and irregularity on the bevel due to incorrect cutting parameters. If laminations are found, they must be investigated in detail per applicable specifications. In the case of irregular cut edges, the irregularities shall be removed by buttering and subsequent dressing. However, the buttering shall be done only on a clean surface that is free from the remains of the oxyacetylene/plasma arc cutting. The plate is then prepinged at both ends first and then completely bent to the required diameter in a plate bending machine. Depending on the amount of curvature needed, the plate is passed through the roller a number of times, to achieve the curvature in stages so that any elongation to the plate will be negligible. The various parameters of the shell section, like the diameter, profile, out of roundness, etc., are checked and if found satisfactory, the shell is taken for longitudinal seam fit up.

5.4.2 Shells Made from Multiple Plates

If the length of plate is not sufficient for making the shell, two or more plates are joined together to obtain the required shell length. This joining is done in plate-to-plate fashion. As there is a possibility for the squareness to be lost while joining, the plate shall be checked for squareness after the full shell length is marked on the joined plate, and if needed, correction by trimming is performed.

The plate shall be bent only after completing the weld from both sides. The reinforcement shall be ground off to the parent metal level to facilitate the smooth bending of the plate in the machine. In case the seams are to be radiographed, it is to be done after bending.

5.5 Cones or Reducers

Cones are produced by two methods. When the bevel angle is comparatively small and the diameter is comparatively large, it can be made in a plate bending machine by independently adjusting the pressure rollers at both ends of the bending machine. This work needs expert operators compared to shell bending. In some cases, the cone has to be made in one piece. This depends mainly on the available plate size, machine capacity, etc.

When the thickness of the cone is larger and the diameter is smaller, cones are made by pressing. The pressing is carried out using matching male and female dies either in full length or in pieces depending on the machine capacity. In both methods, the manufacture of eccentric cones is much more difficult than concentric cones and hence they are made in two halves, provided the machine capacity permits.

6 MEASUREMENT OF DIMENSIONS

6.1 Flat Ends

6.1.1 Dimensions to Be Checked

The main dimensions that are to be checked are the diameter and thickness of the flat head. Being a flat surface, obviously the flatness also has to be checked. Apart from these, there is nothing else to be checked for a flat head.

6.1.2 Methodology

Depending on the size of the head, the diameter of the head can be measured by a standard steel measuring tape and is to be taken at least at 0, 45, 90, and 135° (in four directions). The straightness can be checked using a straight edge and feeler gauge combination for diameters below 1000 mm. Beyond this dimension, the use of a straight edge is not recommended due to unavailability of large straight edges. In that case the use of a stretched piano wire and taper gauge combination as described herein is preferred. The piano wire has to be stretched over the surface of the flat head over standard blocks of known dimensions. Check the taper gauge reading between the stretched piano wire and the flat head at fixed intervals. The deviation in reading from the standard block thickness is the out-of-straightness. In addition to this, the thickness at various locations has to be checked by using a micrometer at accessible locations as well as by using a calibrated ultrasonic thickness gauge. The dimensions mentioned shall be recorded in the dimension report that accompanies the head. Figure 3.6 depicts the methodology.

6.2 Dished Ends

6.2.1 Dimensions to Be Checked

The following dimensions shall be recorded, as shown in Figure 3.7.

1. Actual inside diameter at a minimum of four locations.
2. Outside circumference at straight flange (SF) and thickness at SF from at least 12 locations.
3. Depth of dish (inside) excluding SF and SF.
4. Thickness reading at 100-mm spacing along four lines radiating from the center of the head to SF at 0, 90, 180, and 270° marked on the outside surface at SF using an ultrasonic thickness gauge (UTG).

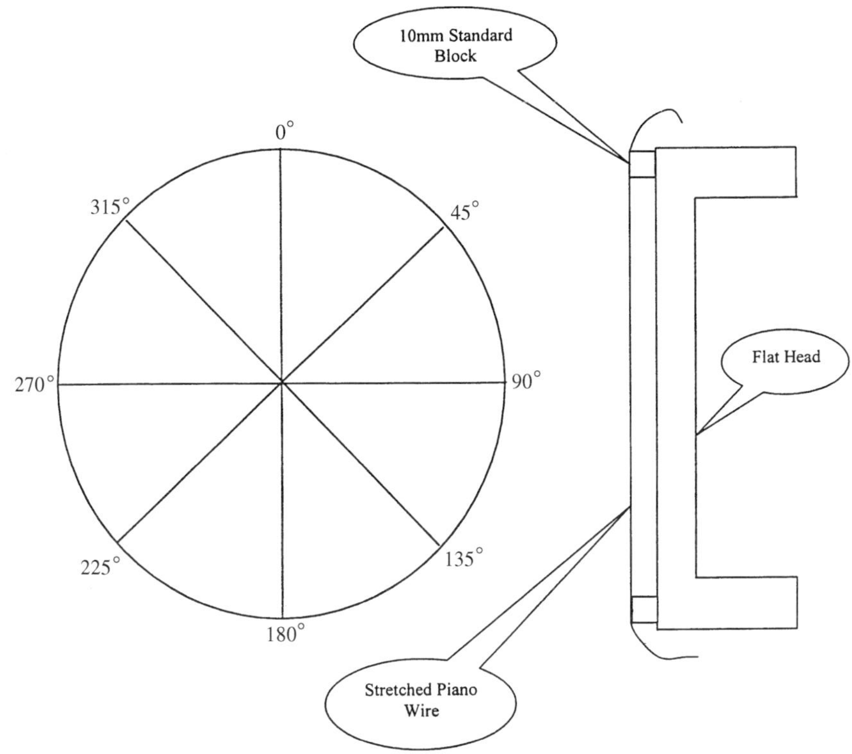

Figure 3.6 Dimensional check of flat heads.

Whenever a UTG is used for measuring the thickness, the readings shall be furnished in the format depicted in Figure 3.8.

6.2.2 Methodology

Upon manufacture of a dished end as required in the specification, the inside diameter (ID) at four directions shall be recorded using a measuring tape. Similarly, the outside circumference at the straight face also shall be recorded. The thickness at SF at various locations can be taken accurately using a micrometer. This can also be taken using the UTG

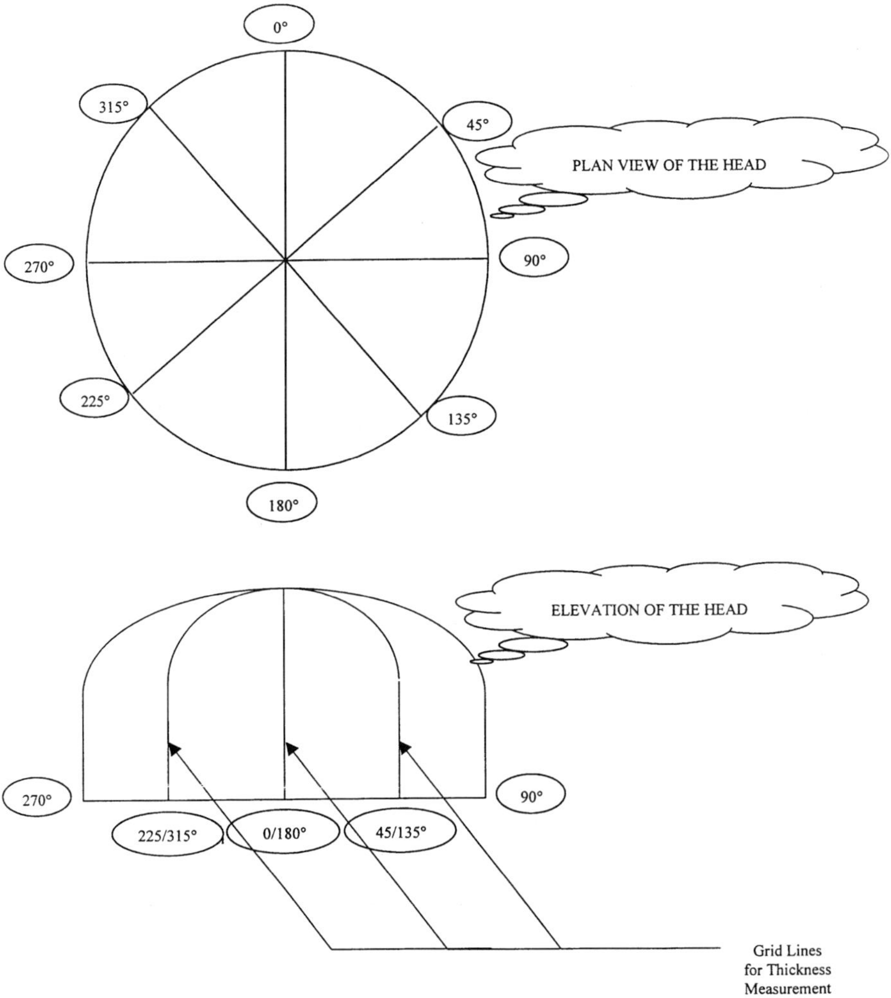

Figure 3.7 Dimensional check of dished ends.

EMBLEM

ULTRASONIC THICKNESS GAUGING REPORT

Order No.		Sheet	of
Purchaser		Attachment to Certificate No	
Description		Date	
Drawing No(s)			

Test Equipment		Calibration Blocks	
Test Method		Couplant	
Probe		Material under Test	
Measuring accuracy		Surface Condition	

Sl.No.	Identification	Thickness (mm)				Remarks
		T1	T2	T3	T(av)	

Figure 3.8 Ultrasonic thickness gauging report.

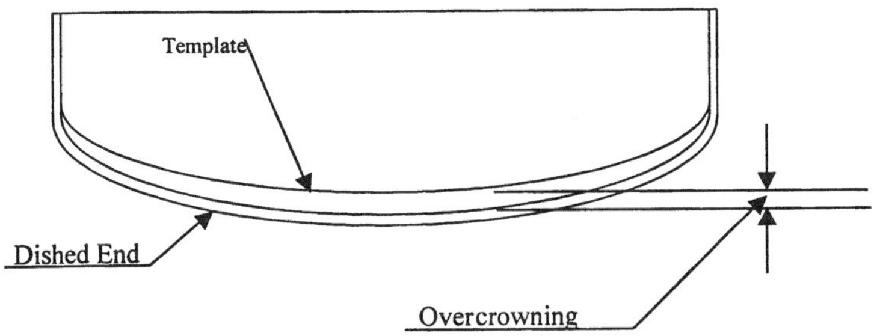

Figure 3.9 Overcrowning of dished ends. (*Note*: The overcrowning should be checked in at least four directions and the maximum value recorded.)

but wherever possible, it is recommended to use a micrometer, which is very simple in operation. The overcrowning as well as the out-of-profile is checked by the use of a template (made per the design) and a taper gauge. Figure 3.9 may be referred to for details.

6.3 Shells

6.3.1 Dimensions to Be Recorded

The following dimensions shall be recorded without fail:

1. Outside circumference at both ends as well as at the center of the shell
2. Inside or outside diameter (OD) at four angles on either side of the shell
3. Straightness of the shell through center lines at 0, 90, 180, and 270°C
4. Length of shell at four center lines

6.3.2 Methodology

Bend the shell and tack weld the longitudinal seam as required. In cases of thin shells with large diameter, provide spiders (at least four numbers) to maintain the circularity. Measure the circumference at both the edges and at the center of the shell by tightly stretching a steel measuring tape.

The ID at four angles should be taken using steel tape after providing the stiffeners at both ends as shown in Figure 3.10.

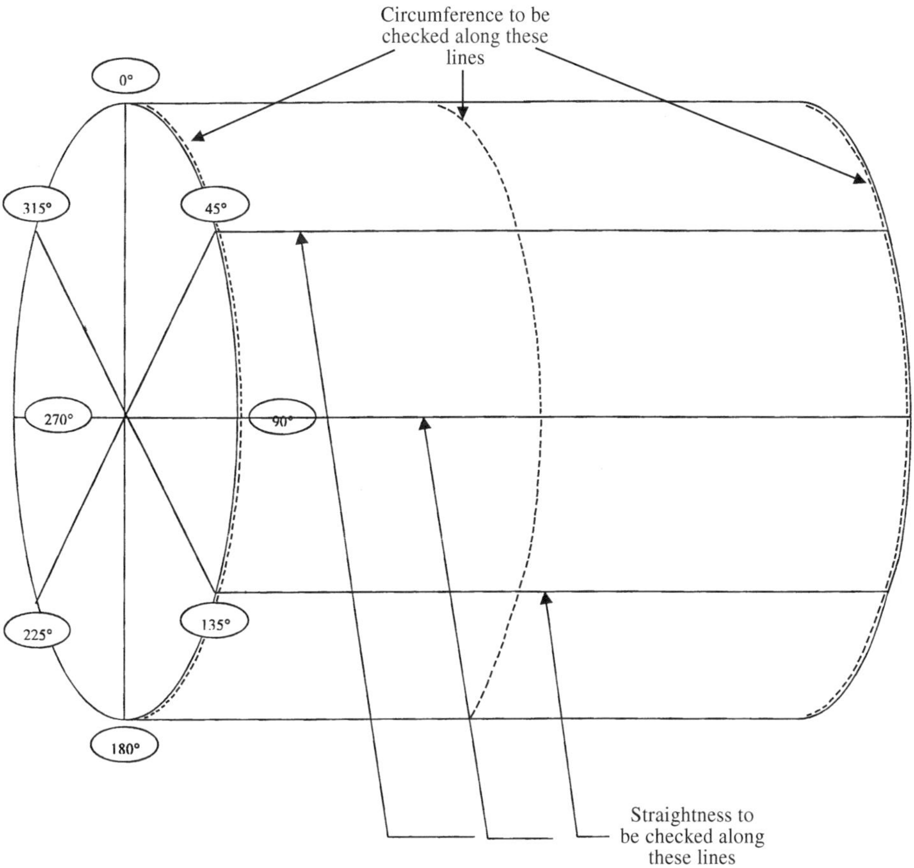

Figure 3.10 Dimensional check of shells.

Straightness of the shell is to be measured and recorded using a stretched piano wire over standard machined blocks of 10- or 20-mm thickness using a taper gauge.

The length of the shell at four locations over the circumference is to be measured and recorded using steel measuring tape.

The profile of the shell should be examined over the entire area, either inside or outside, using the template as specified in UG 29.2. The deviation from the designed profile shall be measured by inserting a taper gauge between the template and the shell. The templates can be either inside or outside, as shown in Figure 3.11A and B.

6.4 Cones or Reducers

6.4.1 Concentric

Dimensions. The following dimensions shall be measured:

1. Outside circumference at both the ends
2. Inside or outside diameter at both ends
3. Slant height of the cone at four diametrically opposite points
4. Straightness of the cone in cases of larger slant height or when circumferencial seams are unavoidable
5. Thickness at random points
6. Profile at both ends
7. Concentricity

Methodology. Bend the cone and tack weld the longitudinal seams and put it on rotators so that it can be turned freely. Measure the outside circumference at both ends at the locations shown in Figure 3.12 in the farthest point possible by stretching the steel tape very tightly to the shell and maintaining the position of the tape exactly in one plane. Due to the configurational deviation at the edges due to edge preparation, there could be a little mismatch in the measured circumfer-

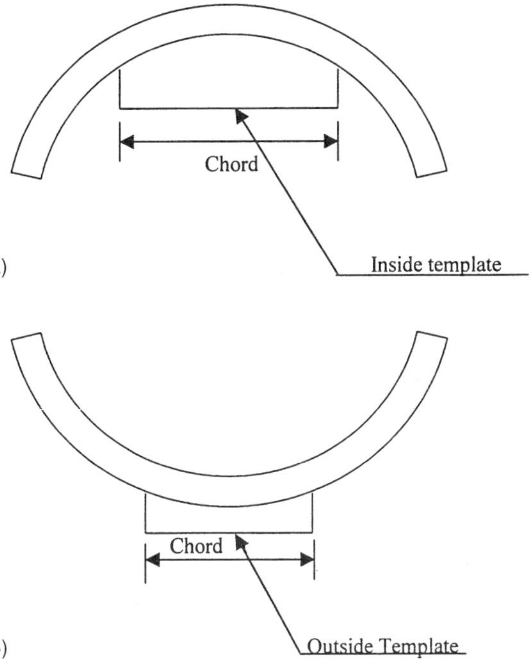

(A)

(B)

Figure 3.11 Templates: (A) inside; (B) outside.

ence based on (ID + $2t$) π. However, this will be negligible compared to the fabrication tolerance provided in the code.

Also measure the ID or OD at both ends at four angles, as done in the case of shells.

The thickness of the cone can be taken at the edges using a micrometer. If there is no noticeable thickness reduction, then it need not be recorded. Otherwise it is to be plotted along four lines connecting the four center lines at both ends.

6.4.2 Eccentric

Dimensions. The following dimensions shall be measured:

1. Outside circumference at both the ends
2. Inside or outside diameter at both ends

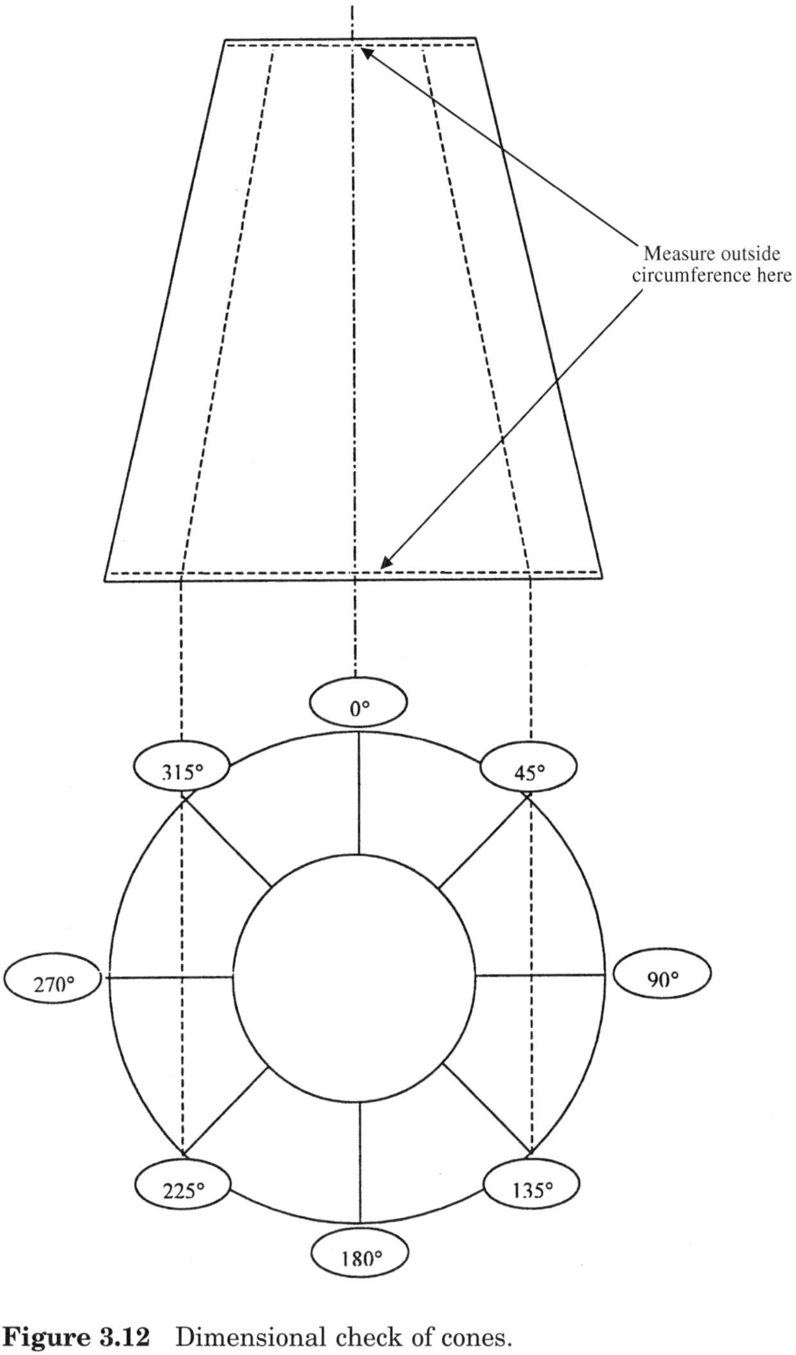

Figure 3.12 Dimensional check of cones.

3. Slant height of the cone
4. Straightness of the cone in cases of larger slant height or when circumferencial seams are unavoidable
5. Thickness at random points
6. Profile at both ends
7. Eccentricity

Methodology. Same as for concentric cones and reducers.

7 INSPECTION

7.1 Dished Ends

1. Dimensions discussed in Sections 6.1.1 and 6.2.1 and subsequently recorded in the related certificate must be physically counterchecked in cases of bought-out items.
2. Profile of the dished end all around shall be checked using a template, and local deformations need to be quantified, including overcrowning.
3. If edge preparation is mentioned in the order, the LPT of the prepared edge shall be carried out.
4. Visual inspection of the surface (especially the outside) and LPT of the crown, knuckle radius portion, and the SF is to be carried out without fail.
5. Thickness as recorded may be counterchecked.
6. When dished ends are seam welded prior to forming, reports to all the specified tests (destructive and nondestructive) are to be verified for compliance. The tolerance for the formed end shall be per UG 81 of ASME Section VIII Div (1) (see Chapter 21, Section 2 of this volume).

7.2 Shells

1. All dimensions mentioned in Section 6.3.1 must be verified.

2. Edge preparation must be inspected for its compliance to that specified in the drawing, such as the V configuration, angle of bevel, root face, root gap, and the uniformity of these aspects throughout the length of the seam.
3. Check the profile with a template whose cord subtends an angle of at least 20° at the center. When the vessel is expected to take external pressure, the maximum plus or minus deviation from the true circular form measured radially on the outside or inside circumference shall not exceed the maximum value of e obtained from UG 80.1 (see Chapter 21, Section 3). Use $e = 1.0t$ or $e = 0.2t$, respectively, for points falling above and below the curves. The templates used for such measurement shall have dimensions as per UG 29.2 (see Chapter 21, Section 4). This has to be carried out throughout the circumference as well as the length of the shell.

   For example, consider the case of a pressure vessel of tan to tan length 8000 mm, 2000 mm outside diameter, and thickness 20 mm. $OD/t = 2000/20 = 100$, and $L/OD = 8000/2000 = 4$. This corresponds to an approximate value of $e = 0.75t = 0.75 \times 20 = 15$ mm. Similarly, for the size of the template, for $OD/t = 100$ and $L/OD = 4$, the arc length of the template $= 0.300$ OD $= 0.300 \times 2000 = 600$ mm.
4. Check the difference between the maximum and minimum ID and ensure this is less than 1% of the ID. When the cross-section passes through an opening or within one ID of the opening, the limit may be increased by 2% of ID of the opening, i.e., tolerance = 1% of ID of the vessel + 2% of the ID of the opening. For vessels with longitudinal lap joints, the permissible tolerance may be increased by the nominal thickness of the plate.
5. Check the outside circumference and ensure that the same is within ±0.25% of the calculated outside cir-

cumference based on the actual thickness of the plate used. (This tolerance is taken from BS 5500, as there is no mention about it in ASME.) However, this limit is applicable only for the first component of the vessel, as the individual shells are expected to match each other. This restriction on tolerance due to other specifications is explained with the help of an example.

The tolerance for alignment of shells is given in UW 33 of ASME Section VIII Div (1) (see Chapter 21, Section 5). According to this, for a shell thickness of 10 mm (for category A, B, C, and D welds), the permitted mismatch between the shells is $\frac{1}{4}\ t = 2.5$ mm. Assume the worst case, where there is 2.5 mm mismatch all around the circumference. In that case, the second shell can be larger or smaller in circumference by $5\pi = 15.70$ mm. In other words, to meet the alignment tolerance of the code, the maximum variation in circumference between shells is only ± 15.70 mm. Because of this restriction, the maximum circumferential tolerance of $\pm 0.25\%$ can only be availed by the manufacturer who makes the first component; usually it will be for the dish. For the very same reason, it is recommended to begin work on shells only after the receipt of the dished ends.

7.3 Cones or Reducers

1. All dimensions as delineated in Section 6.4 shall be verified.
2. Apart from these, the concentricity/eccentricity of the cone must be checked as follows:

   Step 1. Make a level base plate and draw the OD at the big end and the ID at the small end per the design on the base plate.

   Step 2. Put gusset plates all around the small end OD and place the cone inside exactly.

   Step 3. Drop plumb from the OD at the big end from

at least eight different points, namely, 0, 45, 90, 135, 180, 225, 270, and 315° after leveling the big end using a spirit level. The plumb should coincide with the OD already marked on the base plate. If not, measure the maximum deviation from the designed shape and it should be well within that specified in the code.

Step 4. Check the entire outside surface of the cone for any local deformations or pressmarks. If seen, these shall be removed by grinding, and LPT may be carried out to rule out surface defects.

Step 5. Check both top and bottom of the cone using templates of the required diameter for compliance. Deviation, if any, shall be recorded and must be within the tolerances specified as in the case of shells.

7.4 Acceptance Norms

For all shapes, objectionable flat areas are to be avoided. For shells under internal pressure, the shell shall be substantially round. Further, the difference between maximum and minimum ID shall be as mentioned in Section 7.2.

8 DESTRUCTIVE AND NONDESTRUCTIVE TESTING REQUIREMENTS

8.1 General

Destructive and nondestructive testing (DT and NDT) requirements are framed based on the code requirements and also on the basis of specific requirements for the project. Usually this will be consolidated in the drawing by the design department as well as by the planning department in their work order released to the production. For carrying out the DT/

NDT, the relevant sections of ASME Sections VIII and V are to be followed as needed and must be closely monitored by the QAC department.

8.2 Destructive Testing

The usual destructive testing envisaged by the code is impact testing. As far as this book is concerned, we will focus on the methodology of these tests. A minimum of three tests shall be performed and the average value may be considered for acceptance. These tests are to be carried out on specimens taken from weld, heat-affected zone (HAZ), and base metal. The temperature at which the test shall be carried out shall be given in the drawing and a suitable bath may used to generate the required temperature. Up to $-50°C$, dry ice is very effective. As there is no positive record of these tests, it shall be done in the presence of the concerned engineer or the third party inspector as needed.

As far as possible, the test may be carried out using standard specimens of 10 mm $\times$ 10 mm cross-section. Smaller specimens may be used only when the thickness of the welded part is below 10 mm. The acceptance of these test results shall be as described in UG 84. In the case of alloy steel materials, the applicable section is UHA 51.

8.3 Nondestructive Testing

The most commonly deployed method of NDT in manufacture of pressure vessel is radiographic testing (RT); the rest of the methods put together form only a minor segment of total NDT. There are only three categories of NDT requirement, namely, full, spot, and nil. This decision basically depends on the service, thickness, and joint efficiency used in the design. Therefore, by the time the drawing is prepared and accepted, the category will be decided. For implementation procedures, refer to UW 11, UW 50, UW 51, and UW 52 (see Chapter 21, Sections 6–9).

The gist of the stated clauses is as follows:

1. For full RT, all butt welds in shell and head have to be fully radiographed.
2. Circumferential seam welds on nozzles below 10 in. need not be radiographed if the thickness of the nozzle is less than 1$^1/_8$ in.
3. For spot RT, one spot on each vessel is tested for each 50-ft increment of the weld or fraction thereof. For nozzles the very same condition for full radiography shall apply.
4. The acceptance criteria for all radiographs shall be per UW 51 and UW 52 for linear defects and Appendix 4 for rounded indications [Appendix 4 of ASME Section VIII Div (1) (see Chapter 21, Section 10)].

From the applicable specification UW 11, ultrasonic examination carried out in accordance with UW 53, (see Chapter 21, Section 11) may be substituted for radiography for the final closure seam of a pressure vessel if the construction of the vessel does not permit interpretable radiographs. The technique adopted shall be in accordance with Appendix 12 of ASME Section VIII Div (1) (see Chapter 21, Section 12). For acceptance criteria also, the same section shall be referred to.

For other NDT methods like MPT and LPT, the applicable sections are UG 93 and UHT 57 respectively. For both these NDT methods, the basic requirements on methodology evaluation and acceptance shall be per Appendices 6 and 8 of ASME Section VIII Div (1), respectively (see Chapter 21, Sections 13 and 14).

8.4 Records

For all DT, reports must be generated and countersigned by all the parties who witness the test. If required by the client, the DT will consist of all weld tensile, transverse tensile, and bend tests; charpy or izod impact hardness test; and macro-

EMBLEM

MECHANICAL TEST REPORT

Order No.		Sheet	of
Purchaser		Attachment to Certificate No	
Description		Quantity	
Drawing No(s)		Date	

1 Reduced Section Tensile Test

Specimen No.	Dimensions (L x B) mm	Area (mm^2)	Gauge Length (mm)	UTS (kg/mm^2)	YP (kg/mm^2)	Elongation (%)	Remarks

2 All Weld Metal Tensile Test

Specimen No.	Dimension (diameter) mm	Area (mm^2)	Gauge Length (mm)	UTS (kg/mm^2)	YP (kg/mm^2)	Elongation (%)	Remarks

3 Guided Face Bend Test

Specimen No.	Thickness of Sample (t) mm	Former Radius	Remarks

4 Guided Root Bend Test

Specimen No.	Thickness of Sample (t) mm	Former Radius	Remarks

Figure 3.13 Mechanical test report.

5 **Guided Side Bend Test**

Specimen No.	Thickness of Sample (t) mm	Former Radius	Remarks

6 **Nick Break Test**

Specimen No.	Type of Nick	Specimen Thickness (mm)	Remarks

7 **Impact Test**

Specimen No.	Type of Notch	Size (mm)	Test Temperature (° C)	Impact value (kg_f-m)	Remarks

8 **Hardness Test**

Specimen No.	Hardness Value (Brinnel)	Hardness Value (Rockwell)	Hardness Value (Vickers)

9 **Macro Test**

Specimen No.	Type of Etchant	Observation

Figure 3.13 Continued

examination. In almost all the cases the requirement will be only for impact testing at a low temperature. A typical format for this purpose is given in Figure 3.13.

Similarly every RT taken under a particular technique is to be recorded in a report form (Figures 3.14 and 3.15) along with the findings. These will be reviewed by a qualified and

EMBLEM

RADIOGRAPHIC EXAMINATION REPORT

Order No.		Sheet	of
Purchaser		Attachment to Certificate No	
Description		Date	
Drawing No(s)			

Material		Type of Joint	
Welding Process		Surface condition	
Stage of Test		Extent of Test	

X- Ray		Gamma Ray		Material Thickness	
				Film	
kV		Isotope		IQI	
mA		Strength		IQI Placement	
Exposure Time		Exposure Time		Lead Screen: Front	
				: Back	
SFD		SFD		Density	
Focal Spot		Source Size		Sensitivity	

SKETCH

Figure 3.14 Radiographic examination report.

EMBLEM

RADIOGRAPHIC EXAMINATION REPORT- ANNEXURE

Attachment to Certificate No.		Sheet of

Sl.No.	Identification	Date	Film Size	Observation	Remarks

Figure 3.15 Annexure to radiographic examination report.

EMBLEM

ULTRASONIC EXAMINATION REPORT

Order No.		Sheet	of
Purchaser		Attachment to Certificate No	
Description		Date	
Drawing No(s)			

Material		Probe (1)				
Thickness		Probe (2)				
Surface Condition		Couplant				
Direction of Test		Basic Calibration Block				
Extent of Test		Reference Hole		T		Dia
Test method		Reference Level 80 % FSH			dB	+6dB
Test Equipment		Test range				

SKETCH

Figure 3.16 Ultrasonic examination report.

EMBLEM

ULTRASONIC EXAMINATION REPORT- ANNEXURE

Attachment to Certificate No.		Sheet of

Scanning surface	Indications		Direction of Measure-ment	Reading Interval (mm)	Stand Off (mm)	Beam Path (mm)	Type of defect	Remarks
	Sl.No	% Amp						

Figure 3.17 Annexure to ultrasonic examination report.

EMBLEM

MAGNETIC PARTICLE EXAMINATION REPORT

Order No.		Sheet	of
Purchaser		Attachment to Certificate No	
Description		Date	
Drawing No(s)			

Equipment		Surface under Examination	
Magnetizing Process			
Magnetizing Current		Surface Preparation	
Technique		Magnetic Particle	
Direction of Field		De-magnetization	
Field Strength		Post-Cleaning	

Sl.No.	Location	Length of Defect	Remarks	SKETCH

Figure 3.18 Magnetic particle examination report.

EMBLEM

LIQUID PENETRANT EXAMINATION REPORT

Order No.		Sheet	of
Purchaser		Attachment to Certificate No	
Description		Date	
Drawing No(s)			

Code	
Description of Surface	
Temperature of Surface	
Surface Condition	
Pre-Cleaning	
Penetrant Penetrant Application	
Dwell Time Cleaner Removal of Penetrant	
Developer Developer Application	
Developing Time	
Results Post-cleaning	
Remarks	

Figure 3.19 Liquid penetrant examination report.

trained person according to the acceptance criteria, which again will be recorded in the report and will form a part of the final QAC documentation pertaining to the vessel.

As mentioned in the case of RT, every other nondestructive test (namely, UT, MPT, and LPT) shall be recorded properly for future reference. These reports shall contain all salient aspects of the technique used in the NDT so that in future if this is repeated under the same parameters, the results can be compared, especially in the case of UT. The formats developed for this purpose are found in Figures 3.16 and 3.17 for UT, Figure 3.18 for MPT, and Figure 3.19 for LPT.

4

Manufacture of Vessel

1 GENERAL

In this chapter, the topics are arranged according to the sequence of manufacture of a pressure vessel under the following assumptions.

1. The dished ends and cones are bought from outside with the material supplied by the vendor under proper certification.
2. All plates and piping materials, including but not limited to pipes, elbows, flanges, studs/bolts and nuts, gaskets, etc., are bought-out items from reputed vendors with proper certification.

2 RAW MATERIALS INSPECTION

2.1 General

All raw materials for the manufacture of a pressure vessel are ordered from reputed vendors, and the supply shall be accompanied by proper certification. Upon receipt of the material at the stores, the quality assurance and control (QAC) department will be informed through their intimation note, as in Figure 4.1. In the meantime, the procurement department will forward the relevant documents pertaining to the received item to QAC to facilitate the inspection with a copy of the intimation marked to the planning department for information processing. Until inspection is carried out, the raw materials are kept in the incoming material inspection bay of the stores for all materials other than plates and pipes, which are stored separately in the open yard attached to the stores.

Those items having undergone inspection will have either a green, yellow, or red sticker or tag with the initials of the inspection engineer and the date of inspection as shown in Figure 4.2. Green indicates accepted material, yellow is for items under hold, and red marks rejected items.

Items under hold need clarifications from other departments as to their acceptability, when a discrepancy in supply is observed. This is intimated to the concerned department (planning/procurement/design) in detail by an inspection note, as in Figure 4.3. The matter is to be resolved by the concerned department(s), and if the item is to be accepted QAC should be notified. Until such time, the item will be under hold in an area designated for the purpose.

All accepted materials will have a green sticker implying that they meet all the requirements mentioned in the purchase order (PO) as well as the applicable codes and items-supplied tally with the certificates furnished by the vendor.

Those items that do not meet the PO or code requirements shall be rejected and are given a red sticker. The reasons for the rejection are given in the inspection note to be

EMBLEM

INTIMATION OF MATERIAL RECIEPT

Purchase Order No.		Date	
Vendor		Date of Receipt	

DETAILS OF ITEMS RECEIVED

Sl.No.	Tag No.	Description

Location:
Store Keeper in Charge:
Status of MTC: Received/ not received

Signature of Stores Officer:

Routing:-
Manager (QAC)

Cc: Asst. Manager(Purchase)
Manager(Planning)
Manager(Production)

O/C 1 Master File
O/C 2 PO File

Figure 4.1 Intimation of material receipt.

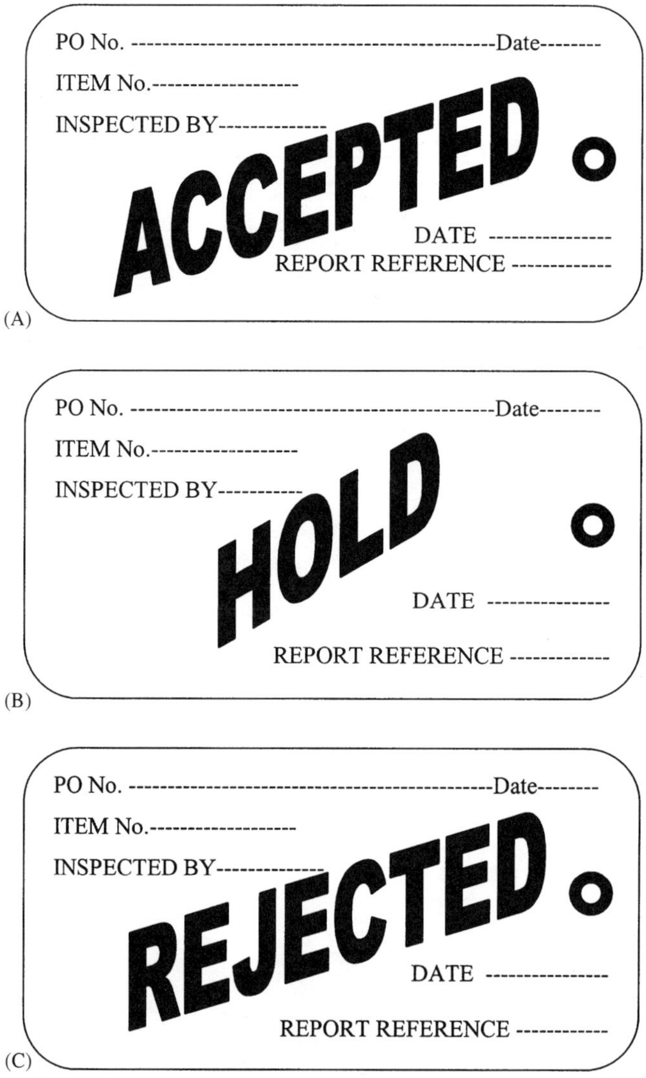

Figure 4.2 Stamps: (A) accepted; (B) hold; (C) rejected.

EMBLEM

STORES INSPECTION REPORT

Purchase Order No.		Date	
Vendor		Date of Inspection	

DETAILS OF ITEMS INSPECTED

Sl.No.	Tag No.	Description	Remarks

In case the items are on hold, please indicate the clarifications needed:

In case the items are rejected, please indicate the reasons:

Signature of Inspection Engineer

Routing:-
Asst. Manager (Purchase)

Cc: Manager (Planning)
Manager (Design & Drawing)
Manager (Production)

O/C 1 Master File
O/C 2 PO File

Figure 4.3 Stores inspection report.

communicated to the vendor by the procurement department, and at the same time the material will be returned to the vendor by the stores department. The procurement department in the meantime must initiate alternative procurement action on a priority basis.

2.2 Dished Ends

Each dish shall be identified against its certificate. A sample certificate for dished end is enclosed as Figure 4.4. Later, the outside circumference, inside diameter (ID) at four locations, thickness at straight flange (SF), and inside depth are measured and cross-checked with the measurements given in the certificate. Further, the heat number and identifications certified by the third party inspector at the manufacturer's end are also verified. If these are all in order, the physical verification part of inspection is over. The certificate is then verified for material specification, types of tests, heat treatment, etc. When all these parameters are verified per the PO and applicable codes, the item is accepted.

2.3 Plates/Pipes

Measure the possible physical dimensions, especially the thickness. Correlate the identification found on these items with the certificate furnished. If these are in order, the certificate is verified against code and PO requirements. If these are also in order, the items can be accepted. Sample certificates for plates and pipes are shown in Figures 4.5 and 4.6, respectively.

2.4 Pipe Fittings/Flanges

Record all possible dimensions and verify against those given in the applicable standards. Cross-check the identification found on the items to that given in the certificate. Verify cer-

MANUFACTURER'S ADDRESS

EMBLEM

CLIENT		INSPECTION AUTHORITY	**LLOYDS REGISTER**
PURCHASE ORDER		CERTIFICATE NUMBER	

ARTICLE — 2:1 ELLIPSOIDAL HEADS
MATERIAL OF HEAD — SA 515-GR 70
ACCORDING TO — ASME Section VIII Div (1)1989 + Add90
STATE OF DELIVERY — COLD PRESSED AND SPUN HEADS NORMALIZED AFTER FORMING
MARKING ON THE HEADS — SPECIFICATION, GRADE, HEAT NO., PLATE NO, HEAD NO & INSPECTOR'S STAMP PNS
INSPECTORS STAMP

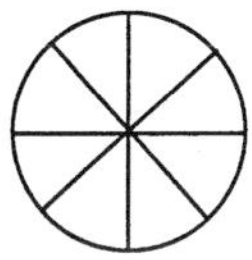

Sl. No.	**Description**	**Dimension Required (mm)**	**Dimension Measured**				**Remarks**
			D1 (mm)	**D2 (mm)**	**D3 (mm)**	**D4 (mm)**	
1	I/S Diameter	1600	1594	1600	1597	1602	
2	Crown radius	1300					
3	Inside Depth	466	478	479	478	480	
4	S F Thickness	20.0	21.3	21.5	21.6	21.4	
5	Knuckle Thickness	15.75	16.5	16.8	16.7	16.8	
6	SF	60	63	65	64	65	
7	KR	250					
8	Overcrowning	20.0	8	10	11	9	
9	O/S Circumference	5154	5145				
10	Inside Surface	Satisfactory					
11	Outside Surface	Satisfactory					
12	D P Test	Satisfactory					
13	Profile	Satisfactory					
14	Heat Treatment	Normalizing after formation					
15	Radiography	N/A					
16	Coupon Plate Testing	N/A					

QA/QC Manager

Figure 4.4 Sample certificate of heads.

MANUFACTURER'S ADDRESS			EMBLEM

CLIENT		INSPECTION AUTHORITY	**LLOYDS REGISTER**
PURCHASE ORDER		CERTIFICATE NUMBER	

ARTICLE	PLATES
MATERIAL	**SA 240-316L / S 31603**
ACCORDING TO	ASME II A Ed92 + Add94
STATE OF DELIVERY	SOLUTION ANNEALED AT 1050/1100°C
MELTING PROCESS	**E**
MARKING	TRADE MARK-GRADE-HEAT NR-PLATE NR-INSPECTOR'S STAMP
BRAND OF MANUFACTURER	ABC
INSPECTORS STAMP	PNS

Extent of Material delivery

Marking	Number	Description	Plate No(s)	Heat No.	Test No.	Weight(kg)
4-9	2	4600x1550x10.0	211268/1-2	51051	211268	1141
3-8	1	8050x1550x10.0	211268/3-3	51051	211268	998
13-13	1	5000x1500x22.0	213282/1-1	51051	213282	1320
8-4	1	5450x1930x12.0	213283/1-1	51051	213283	1010
9-5	1	4400x1930x12.0	213283/2-2	51051	213283	815
Total	6					5283

We hereby certify that the above mentioned products are consistent with the order prescriptions

Date Quality Control manager Mill Inspector

Figure 4.5 Sample certificate of plate.

CHEMICAL ANALYSIS

HEAT	CHEMICAL COMPOSITION %								
	C	Mn	P	S	Si	Ni	Cr	Mo	N
51051	0.018	1.771	0.035	0.0025	0.490	10.309	16.546	2.067	0.0289

MECHANICAL TESTS

Ep(mm)	HEAT	SAMPLE	TYPE	T°C	ETAT	MPa			
						Rp 0.2%	Rp 1%	Rm/ UTS	A150%
10.00	51051	211267	140	20	0	276	319	568	79
22.00	51051	213282	140	20	0	209	241	528	82
12.00	51051	213283	140	20	0	267	296	543	88
						HB30	HB30	HB30	HBMOY
10.00	51051	211267	410	20	0	159	167	166	164
22.00	51051	213282	410	20	0	152	144	145	147
12.00	51051	213283	410	20	0	147	155	154	152

TYPE	140	TRANSVERSE TENSILE TEST	
TYPE	410	BRINNEL HARDNESS	3000kg
ETAT	O	DELIVERY CONDITION	

This certificate contains two pages only

Results of Surface Examination & Dimensions: SATISFACTORY

Date Quality Control manager Mill Inspector

Figure 4.5 Continued

MANUFACTURER'S ADDRESS			EMBLEM

CLIENT		INSPECTION AUTHORITY	**LLOYDS REGISTER**
PURCHASE ORDER		CERTIFICATE NUMBER	

ARTICLE — HIGH ALLOY TUBES-COLD FINISHED

MATERIAL — **SA 312 TP316L**

ACCORDING TO — ASME II A Ed1989 + Add90 IN CONNECTION WITH NACE STANDARD MR-01-75 Rev 90 Section 3.5

STATE OF DELIVERY — SOLUTION ANNEALED AT 1050/1100°C-FREE FROM SCALES-ENDS PLAIN-SQUARE CUT TO TUBE AXIS-DEBURRED-TOLERANCE ACCORDING TO ASTM A 530M-91A-IN RANDOM MILL LENGTH 5000 TO 7000mm

MELTING PROCESS — **E**

MARKING — TRADE MARK-GRADE-HEAT NR-INSPECTOR'S STAMP

BRAND OF MANUFACTURER — ABC

INSPECTORS STAMP — PNS

Extent of Material delivery

Marking	Number	Description	Heat No.	Total Length	Test No.	Weight(kg)
91042	48	1"NB Sch80 33.4 mm OD x 4.55 mm W/T	91042	324.59	-	1076
Total	48					1076

We hereby certify that the above mentioned products are consistent with the order prescriptions

Date Quality Control manager Mill Inspector

Figure 4.6 Sample certificate of pipe.

CHEMICAL ANALYSIS

EAT	CHEMICAL COMPOSITION %								
	C	Mn	P	S	Si	Ni	Cr	Mo	N
91042	0.015	1.720	0.024	0.0020	0.370	12.250	16.740	2.350	0.0289

MECHANICAL TESTS

HEAT NUMBER	TP DIRECTION	YIELD STRENGTH	TENSILE STRENGTH	ELONGATION
91042	Longitudinal	264N/sqmm	599N/sqmm	44%

OTHER TEST RESULTS

Heat Treatment	As per specification
Hydrostatic Test	at 170 Bar--Satisfactory
Flattening Test	Satisfactory
Visual & Dimensional Inspection	Satisfactory
Material identification Test of all tubes by Spectroscopic Analysis	Satisfactory
Inter Granular Corrosion Test According to ASTM A262 Practice E	Satisfactory
HRC Maximum 22 according to NACE MR-01-75	No Objection

The products have been tested in accordance with the purchase specifications and are found to be satisfactory.

Date　　　　Quality Control Manager　　　　Mill Inspector

Figure 4.6 Continued

tificate against PO and code requirements. Sample certificates for pipe fittings and flanges are as shown in Figures 4.7 and 4.8.

2.5 Fasteners and Gaskets

Select random samples and subject them to a dimensional check. Correlate certificates to the identification given to the

MANUFACTURER'S ADDRESS			EMBLEM

CLIENT		INSPECTION AUTHORITY	**LLOYDS REGISTER**
PURCHASE ORDER		CERTIFICATE NUMBER	

Item No	Description	Qty	Material	Lot No	Item status	Mfr. Process	Dimensional Std.
3	2" Sch 80 90° LR Elbow	01	SA403 Gr WP 316L 1992(A93)	6E03/S-715	C	D/F	ANSI B 16.9

CHEMICAL PROPERTIES

Heat	Lot	CHEMICAL COMPOSITION										
		C	Si	Mn	S	P	Cr	Ni	Mo	V	Nb	Cu
6E03/S-715	MTC	0.016	0.270	1.49	0.026	0.040	16.45	12.05	2.19			
	Chk. Test	0.019	0.300	1.44	0.025	0.036	16.40	11.98	2.21			

MECHANICAL PROPERTIES

Heat	Lot	TENSILE TEST				BHN	HEAT TREATMENT
		YP kg/sqmm	UTS kg/sqmm	%RA	%EL		
6E03/S-715		340.80	532.88	64.18			Solution Annealed at 1090°C

Figure 4.7 Sample certificate of pipe fittings.

OTHER TESTS

Ultrasonic Test	N/A
Magnetic particle Test	N/A
Dye Penetrant test	N/A
Bend test	N/A

IMPACT TEST RESULTS

Lot	Test Method	Temp	Absorbed Energy(kg-m)			
			(1)	(2)	(3)	Average

MELTING PROCESS **E**

MARKING TRADE MARK-GRADE-HEAT NR-INSPECTOR'S STAMP

Material as per ASME Section II Part A 1992 inclusive of addenda 1993

PHYSICAL TESTS WITNESSED BY

MANUFACTURERS DECLARED CHEMICAL ANALYSIS VERIFIED AGAINST SPECIFICATION REQUIREMENTS

BRAND OF MANUFACTURER ABC

INSPECTORS STAMP PNS

The products have been tested in accordance with the purchase specifications and are found to be satisfactory.

Date Quality Control manager Mill Inspector

Figure 4.7 Continued

MANUFACTURER'S ADDRESS			EMBLEM

CLIENT		INSPECTION AUTHORITY	**LLOYDS REGISTER**
PURCHASE ORDER		CERTIFICATE NUMBER	

Item No	Description	Qty	Material	Lot No	Item status	Mfr. Process	Dimensional Std.
3	2" Sch 80 WNRF #150	1	SA 105	J-616	C	D/F	ANSI B 16.5

CHEMICAL PROPERTIES

Heat	Lot	CHEMICAL COMPOSITION										
		C	Si	Mn	S	P	Cr	Ni	Mo	V	Nb	Cu
601	J616	0.215	0.250	0.830	0.023	0.021	0.020	< 0.02	< 0.02	< 0.02	< 0.02	< 0.02

MECHANICAL PROPERTIES

Heat	Lot	TENSILE TEST				BHN	HEAT TREATMENT
		YP MPa	UTS MPa	%RA	%EL		
601	J616	340.80	532.88	64.18	30.00	156	N-900°C

OTHER TESTS

Ultrasonic Test	N/A
Magnetic particle Test	N/A
Dye Penetrant test	N/A
Bend test	N/A

Figure 4.8 Sample certificate of flanges.

IMPACT TEST RESULTS

Lot	Test Method	Temp	Absorbed Energy(kg-m)			
			(1)	(2)	(3)	Average

MELTING PROCESS **E**

MARKING TRADE MARK-GRADE-HEAT NR-INSPECTOR'S STAMP

Material as per ASME Section II Part A 1992 inclusive of addenda 1993

PHYSICAL TESTS WITNESSED BY

MANUFACTURERS DECLARED CHEMICAL ANALYSIS VERIFIED AGAINST SPECIFICATION REQUIREMENTS

BRAND OF MANUFACTURER ABC

INSPECTORS STAMP PNS

The products have been tested in accordance with the purchase specifications and are found to be satisfactory.

Date Quality Control manager Mill Inspector

Figure 4.8 Continued

lot. Then check the certificates for compliance to PO and code requirements. A sample certificate for fasteners is shown in Figure 4.9 and one for gaskets is shown in Figure 4.10.

2.6 Ready-Made Cans

Identify the shell and correlate it to the certificate. Measure the outside circumference at both ends, the length of the shell, as well as its thickness at a few locations, as mentioned in Chapter 3, Section 7.2. These are to be cross-checked with

MANUFACTURER'S ADDRESS			EMBLEM

CLIENT		INSPECTION AUTHORITY	**LLOYDS REGISTER**
PURCHASE ORDER		CERTIFICATE NUMBER	

Item No	Description	Qty	Material	Lot No	Item status
3	1 1/4" UNC x 175mm	80	Stud SA 193 Gr B7	L625	C
4	1 1/4" UNC (H=D)	160	Nut SA 194 Gr 2H	M532	C
21	1/2" UNC x 70mm	86	Stud SA 193 Gr B7	L536	C
22	1/2" UNC (H=D)	176	Nut SA 194 Gr 2H	M504	C

CHEMICAL PROPERTIES

Heat	Lot	CHEMICAL COMPOSITION							
		C	Si	Mn	S	P	Cr	Ni	Mo
L625		0.360	0.240	0.650	0.019	0.018	1.020		
M532		0.370	0.250	0.620	0.020	0.017	1.021		
L536		0.360	0.235	0.700	0.021	0.018	1.020		
M504		0.380	0.235	0.720	0.020	0.018	1.021		

MECHANICAL PROPERTIES

Heat	Yield Stress kg/sqmm	UTS kg/sqmm	Elongation %	Reduction in Area %	Heat Treatment
L625 L536	101.56 103.45	110.34 112.54	21.23 22.34	53.42 54.12	Studs quenched at 860°C & tempered at 650°C

Figure 4.9 Sample certificate of fasteners.

Heat	Proof Load Applied Kg	Hardness HRC	Heat treatment
M532 M504	18060 26500	26-36 27-31	Nuts quenched at 860°C & tempered at 650°C

OTHER TESTS

Ultrasonic Test	N/A
Magnetic particle Test	N/A
Dye Penetrant test	N/A
Bend test	N/A

IMPACT TEST RESULTS

Lot	Test Method	Temp	Absorbed Energy(kg-m)			
			(1)	(2)	(3)	Average

MELTING PROCESS **E**

MARKING B7 and 2H respectively on studs and nuts

Material as per ASME Section II Part A 1992 inclusive of addenda 1993

PHYSICAL TESTS WITNESSED BY

MANUFACTURER'S DECLARED CHEMICAL ANALYSIS VERIFIED AGAINST SPECIFICATION REQUIREMENTS

The products have been tested in accordance with the purchase specifications and are found to be satisfactory.

Date Quality Control manager Mill Inspector

Figure 4.9 Continued

MANUFACTURER'S ADDRESS			EMBLEM
CLIENT		INSPECTION AUTHORITY	**LLOYDS REGISTER**
PURCHASE ORDER		CERTIFICATE NUMBER	

GASKETS

Item No	Description	Size	Rating	Quantity	Item status
1		24"	600Psi	02	C
3	Monel Spiral wound	8"	600Psi	04	C
4	asbestos filled gaskets	6"	150Psi	03	C

CHEMICAL PROPERTIES

Heat	Lot	CHEMICAL COMPOSITION							
		C	Si	Mn	S	Cu	Ni	Fe	
Monel 0.2mm		0.0360	0.045	0.980	0.0010	32.400	64.400	1.500	

The material confirms to SB 127-61 as far as the chemical composition is concerned.

OTHER TESTS

Visual and dimensional inspection carried out and found satisfactory.

Compressibility, sealability and recovery tests carried out on sample/equivalent gaskets and found satisfactory.

The products have been tested in accordance with the purchase specifications and are found to be satisfactory.

Date Quality Control manager Mill Inspector

Figure 4.10 Sample certificate of gaskets.

Customer	
Inspected by	LLOYDS
Ref.No.	XXX
Description	Monel spiral wound asbestos filled gaskets
Size	24"NB
Rating	#600
Initial gasket thickness	0.194"
Equivalent gasket size	5.75" x 3.5"
Equivalent test load	134.77T
COMPRESSION TEST	
Test load	134.77T
Thickness after compression	0.130'
Thickness after release of load	0.160"
Recovery(Min 0.010")	0.030"
Dial Gauge deflection at eqt.test load	70(Average of 4 readings)
SEALABILITY TEST	
Test medium	Dry Nitrogen
Test pressure	128 kg/sqcm
Test temperature	Ambient
Duration	3minutes
Result	Satisfactory
Engineer in Charge	

Figure 4.10 Continued

those given in the certificates. Further, if nondestructive testing is involved, its compliance and acceptability are also to be verified from the certificate. A sample certificate for rolled cans is as shown in Figure 4.11.

2.7 Cones

Identify the cone to a particular certificate by identifying the actual stamping details on the plates, and check measurements as in Chapter 3, Sections 6.4 and 7.3. Cross-check those dimensions to these given in the certificate. Subsequently verify the certificate for compliance to PO and code requirements. A sample certificate is shown in Figure 4.12.

MANUFACTURER'S ADDRESS			EMBLEM

CLIENT		INSPECTION AUTHORITY	**LLOYDS REGISTER**
PURCHSE ORDER		CERTIFICATE NUMBER	

ARTICLE	ROLLED STEEL CANS
MATERIAL OF SHELL	SA 312 TP316L
ACCORDING TO	ASME Section VIII Div(1)1989 + Add90
STATE OF DELIVERY	CAN WITH LONGITUDINAL SEAM WELDED & NDT SATISFACTORILY COMPLETED
MARKING ON THE SHELL	SPECIFICATION, GRADE, HEAT NO., PLATE NO & INSPECTOR'S STAMP + WELD SEAM IDENTIFICATION NEAR EACH SEAM
INSPECTORS STAMP	PNS

SKETCH OF SHELL

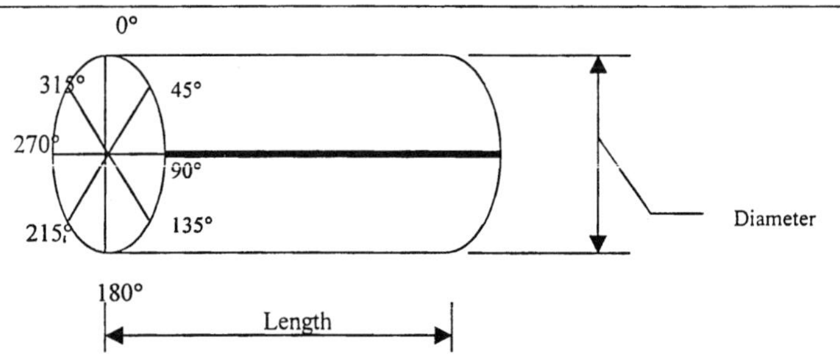

Required Dimensions

DIMENSIONS

Shell Identification	Position Identification	Diameter (mm)av	Length (mm)	Sag/Hog (+ or - mm)
Shell No.1A	0--180°	2505	2500	+3
Shell No.1A	45--225°	2503	2501	+2

Figure 4.11 Sample certificate of shells.

Shell No.1A	90--270°	2495	2499	+4
Shell No.1A	135--315°	2502	2500	+3

Outside circumference (RHS)		Outside circumference (LHS)	
Diameter			
RHS		LHS	
0---180°		0---180°	
45---225°		45---225°	
90---270°		90---270°	
135---315°		135---315°	
Average thickness at RHS		Average thickness at LHS	

OTHER TEST RESULTS

Non-Destructive Testing	
Radiography	100% (Satisfactory)
Ultrasonic	Nil
Magnetic Particle	Nil
Liquid Penetrant	After back gouging/grinding (Satisfactory)
Heat Treatment	As per specification
Hydrostatic Test	Nil
Visual & Dimensional Inspection	Satisfactory
Inter Granular Corrosion Test According to ASTM A262 Practice E	Satisfactory
HRC Maximum 22 according to NACE MR-01-75	Below 20

ENCLOSURES

1) Material test certificate of the plate
2) NDT reports
3) Other test reports

The products have been tested in accordance with the purchase specifications and are found to be satisfactory.

Date Quality Control Manager Mill Inspector

Figure 4.11 Continued

MANUFACTURER'S ADDRESS			EMBLEM

CLIENT		INSPECTION AUTHORITY	**LLOYDS REGISTER**
PURCHSE ORDER		CERTIFICATE NUMBER	

ARTICLE	ROLLED STEEL CONES
MATERIAL OF SHELL	SA 312 TP316L
ACCORDING TO	ASME Section VIII Div(1)1989 + Add90
STATE OF DELIVERY	CONE WITH LONGITUDINAL SEAM WELDED & NDT SATISFACTORILY COMPLETED
MARKING ON THE SHELL	SPECIFICATION, GRADE, HEAT NO., PLATE NO & INSPECTOR'S STAMP + WELD SEAM IDENTIFICATION NEAR EACH SEAM
INSPECTORS STAMP	PNS

SKETCH OF CONE

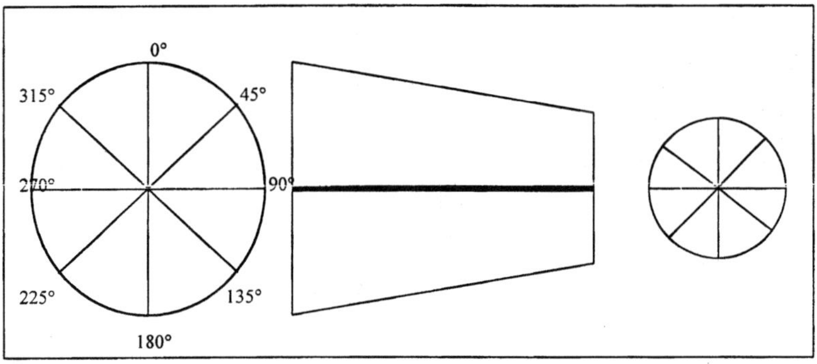

Required Dimensions

DIMENSIONS

Cone Identification	Position Identification	Diameter (mm)av	Slant Height (mm)	Sag/Hog (+ or - mm)
Cone No.1A	0--180°	2505/1802	2500	+3
Cone No.1A	45--225°	2503/1803	2501	+2

Figure 4.12 Sample certificate of cones.

Cone No.1A	90--270°	2495/1800	2499	+4
Cone No.1A	135--315°	2502/1801	2500	+3

Outside circumference (RHS)Big end		Outside circumference (LHS) Small End	
Diameter			
RHS		LHS	
0---180°		0---180°	
45---225°		45---225°	
90---270°		90---270°	
135---315°		135---315°	
Average thickness at RHS		Average thickness at LHS	

OTHER TEST RESULTS

Non-Destructive Testing	
Radiography	100% (Satisfactory)
Ultrasonic	Nil
Magnetic Particle	Nil
Liquid Penetrant	After back gouging/grinding (Satisfactory)
Heat Treatment	As per specification
Hydrostatic Test	Nil
Visual & Dimensional Inspection	Satisfactory
Inter Granular Corrosion Test According to ASTM A262 Practice E	Satisfactory
HRC Maximum 22 according to NACE MR-01-75	Below 20

ENCLOSURES

1) Material test certificate of the plate
2) NDT reports
3) Other test reports

The products have been tested in accordance with the purchase specifications and are found to be satisfactory.

Date Quality Control Manager Mill Inspector

Figure 4.12 Continued

3 MARKING OF SHELLS TO SUIT HEADS AND IDENTIFICATION

This section applies when inside diameter of the shell and dish matches. Take the outside circumference of the dish at straight flange (SF). Measure thickness of the head at SF from at least 12 points and take the average. Calculate the average ID of the dished end. Take the actual thickness of the shell plate. Find the mean diameter of the shell by adding the measured thickness to the calculated ID of the dished end. Based on this, find the mean circumference of the shell. A sample calculation follows:

Required ID = 2500 mm (assuming that the ID of the shell and dish match)
Thickness at SF = 20 mm (per requirement)
Average thickness at SF taken at 12 points (actual) = 21.2 mm
Measured outside circumference of the dish at SF (actual) = 8000 mm

$$\text{Calculated ID of dish} = \frac{8000}{\pi} - 2 \times 21.2 = 2504.07 \text{ mm}$$

$$\approx 2504 \text{ mm}$$

Let the shell thickness required be 16 mm and the actual be 16.5 mm. The calculated mean circumference of the shell is

$$= (2504 + 16.5)\pi$$
$$= 7918.38 \text{ mm} \approx 7918 \text{ mm}$$

Spread out the shell plate and mark the length 7918 mm as shown in Figure 4.13 and cross-check the diagonals D1 and D2. The difference between D1 and D2 shall be within ±2 mm. If this is not possible using the mill cut edge of the plate, the plate must be squared again and the length and breadth remarked. If the diagonals are within the tolerance, the plate identification has to be transferred to a corner of the shell

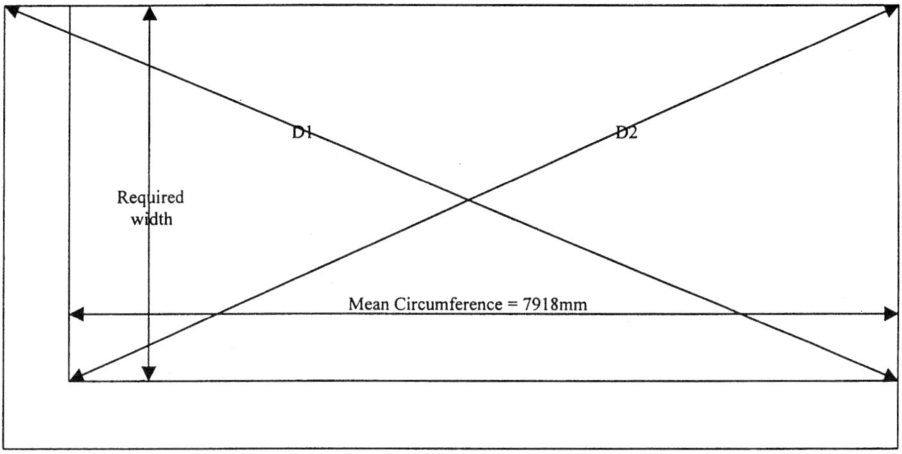

Figure 4.13 Shell marking. The difference between D1 and D2 shall be within ± 2 min.

plate with a personal identification stamp of the inspection engineer who witnesses the transfer of identification. The release for cutting may be issued in the format shown in Figure 4.14 with the signature of the third party inspection agency involved.

4 INSPECTION OF PLATES AFTER CUTTING AND EDGE PREPARATION

The most commonly used process for the cutting and edge preparation of carbon steel plates is oxyacetylene cutting, and that for stainless steel plates is plasma-arc cutting. In both cases, the cut edges will have excessive oxide deposition that is black in color. This is to be removed by grinding, and the acceptable level is at least 1.0 mm deep into the parent metal so that the detrimental effects due to the presence of oxides shall be completely removed.

EMBLEM

CERTIFICATE OF MATERIAL IDENTIFICATION AND TRANSFER OF IDENTIFICATION

Order No.		Sheet	of
Purchaser		Attachment to Certificate No	
Description		Date	
Drawing No			

Part No(s). Marked												

Original Identification Found	Identification Transferred with Drawing No & Part No

SKETCH OF PLATE

Number of scrap pieces identified (with approx. size & shape) :

Material Marked by :

Material Identified by :

Figure 4.14 Certificate of material identification and transfer of identification.

After grind-back, the cut edges shall be examined visually for any defects like lamination or any other local defects due to improper cutting parameters. In cases needed, visual examination of the cut edges may be supplemented by liquid penetrant testing (LPT). When cuts are found satisfactory the piece can be released for bending.

5 BENDING OF PLATES

5.1 Direction of Bending

The shell plate will have more strength in the direction of rolling compared to its transverse direction. Hence, as far as possible, the direction of bending of the plate shall be same as that of the original direction of rolling, which is in the direction of the length of the plate, as shown in Figure 4.15.

5.2 Prepinging

Once the shell plate is fed into the bending machine, first both the ends are pressed to the required shape, as shown in Figure

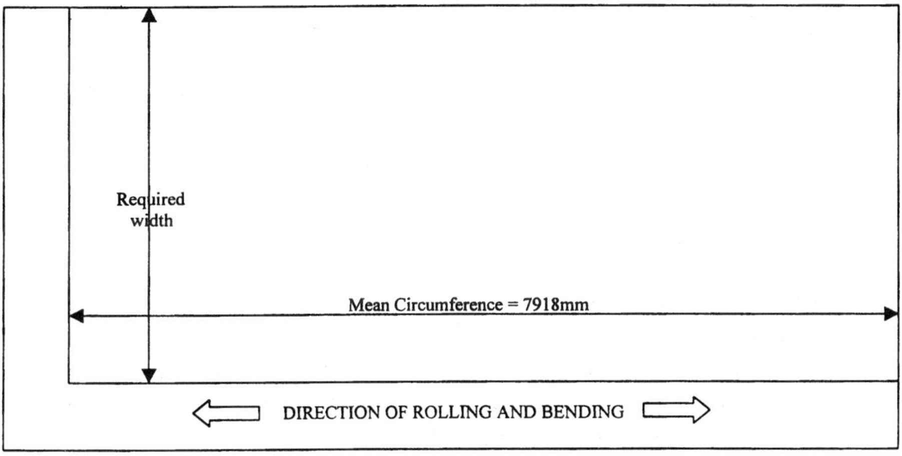

Figure 4.15 Direction of plate rolling and bending.

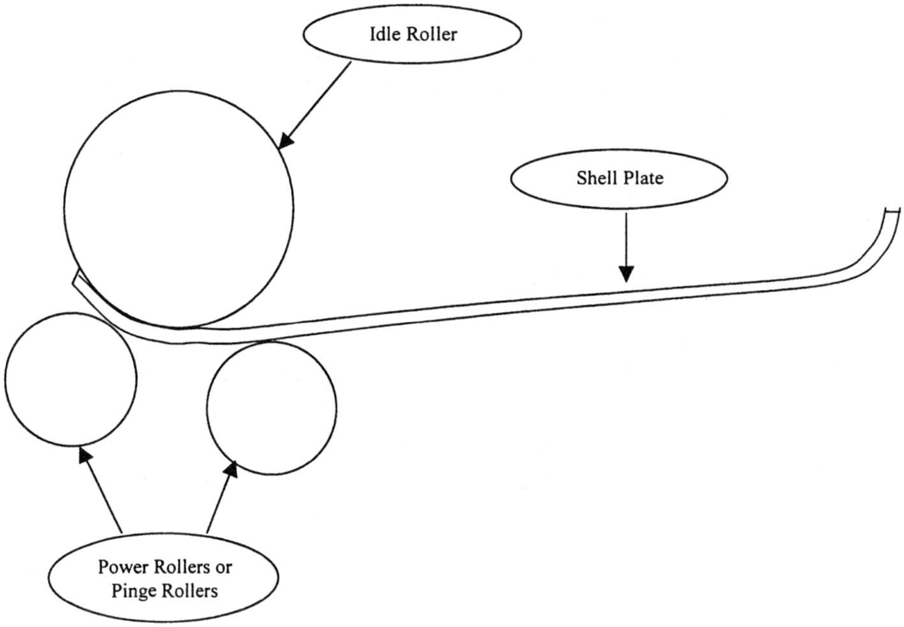

Figure 4.16 Prepinging of plate edges.

4.16. This process is called prepinging. After prepinging, the profile at the ends is checked using a template and, if satisfactory, full bending is carried out. Usually, bending is carried out in a few stages so that the elongation of the shell will be negligible and no further trimming of the plate will be needed.

5.3 Templates

The template for checking the profile during bending shall be as per requirements of UG 29.2 namely, twice the arc length obtained from chart UG 29.2 (see Chapter 21, Section 4). As a rule of thumb, if we take a segmental template that subtends 20° at the center, it will definitely meet the requirements of the ASME clause. The QA/QC engineer shall verify the profile and chord length of the template, and these are to be stamped on the template with the inspection engineer's stamp by the side.

5.4 Set-Up for Welding of Long Seams and Inspection

Upon completion of bending, the longitudinal seam is fitted for welding. Dimensions as listed in Chapter 3, Section 6.3.1 shall be recorded in the format shown in Figure 4.17. Apart from that, the profile all around the shell shall be checked and deviation in profile from the template beyond acceptable levels shall be repaired.

In addition to that, the V configuration shall be checked for its compliance to the design and its uniformity through out the length of the seam (e.g., gap, root face or throat, V angle, etc.).

Further, it would be advisable if one could check the adequacy of preset and restraints given to the seam prior to welding in order to minimize distortion.

5.5 Precautions to Control Distortion

Distortion of weld joints is more problematic in the case of longitudinal seams when compared to the circumferential seams. The usual types of distortion that can take place at a long seam are either peaking in or peaking out. Given here are three proposals for the reduction of distortion whose judicious application will yield positive results.

When the thickness of the shell plate is less than 16 mm, single-V double-welded joints are recommended. In such cases, the groove shall be provided as far as possible at the inside, as shown in Figure 4.18A. When the thickness of the shell plate is greater than 16 mm, double-V joints are recommended. In this case, as far as possible, the joint configuration shall be unequal V with the major V on the inside as shown in Figure 4.18B. While the welding from the inside is being

EMBLEM

SHELL DIMENSION REPORT

Order No.		Drawing No.	
Purchaser		Date	
Description		Status of Shell	

SKETCH OF SHELL

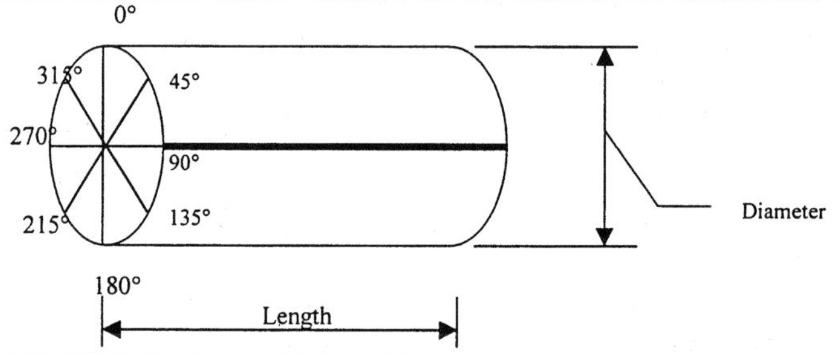

Required Dimensions

DIMENSIONS

Shell Identification	Position Identification	Diameter (mm)	Length (mm)	Sag/Hog (+ or - mm)
Shell No.1A	0--180°	2505	2500	+3
Shell No.1A	45--225°	2503	2501	+2
Shell No.1A	90--270°	2495	2499	+4
Shell No.1A	135--315°	2502	2500	+3

QA/QC Manager

Figure 4.17 Shell dimension report.

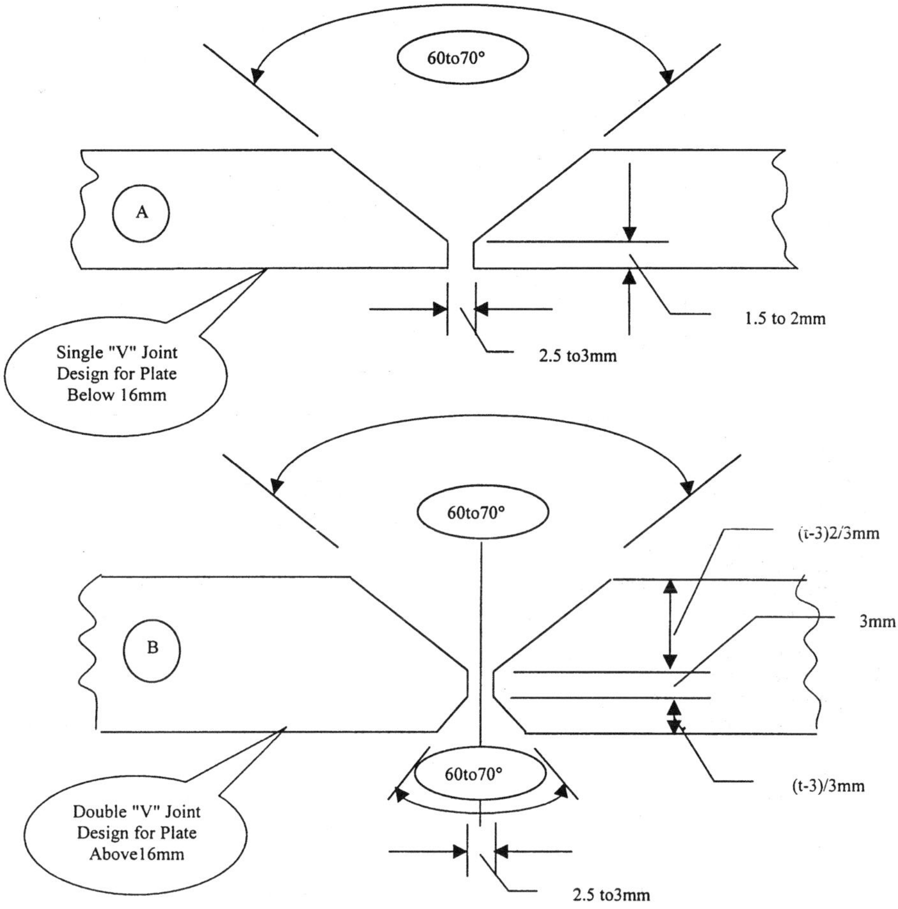

Figure 4.18 Recommended V joint preparation.

carried out, there shall be at least two spiders in position, as shown in Figure 4.19, at both ends of the shell. While providing the spiders, care shall be taken to avoid the weld area to facilitate easy access to the welder.

Upon completion of welding from the inside and prior to the back-gouging, two more stiffeners shall be provided, as shown in Figure 4.20. While providing this, a preset up to D +

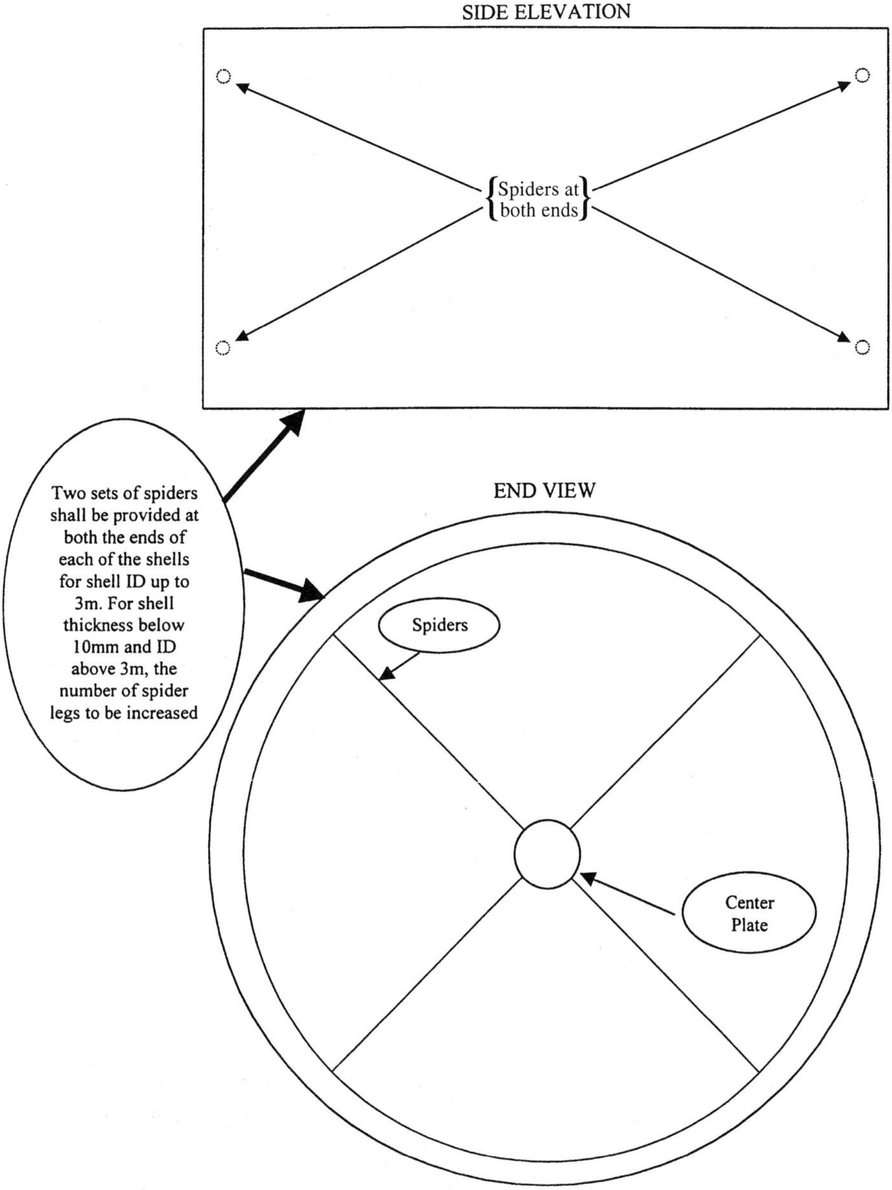

Figure 4.19 Fixtures to maintain circularity.

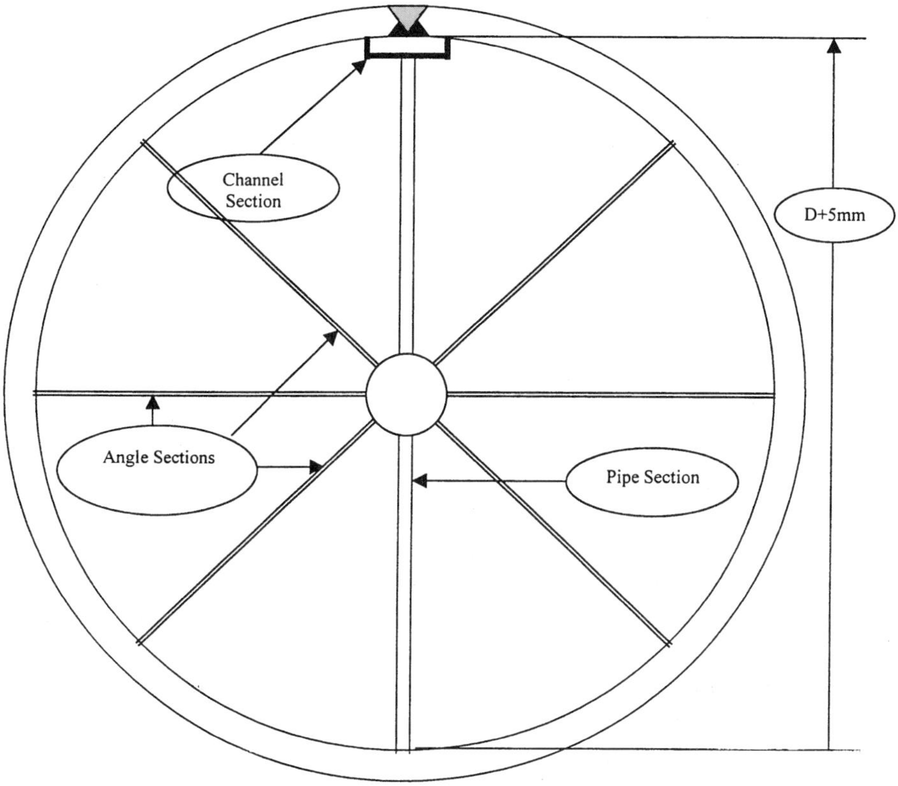

Figure 4.20 Restraint on long seam for welding from outside. *Note*: (1) The channel section shall be tack welded only on one side; (2) the pipe shall be tack welded to the channel at the top and the shell at the bottom; (3) channel and pipe of suitable size and thickness shall be used based on the thickness and diameter of the shell.

5 mm is provided based on the diameter of the shell. This shall be maintained during back-gouging as well as during the completion of welding from outside. The additional channel stiffener provided over the weld seam may be removed prior to the radiographic testing (RT) and the remaining three spiders may be retained until the vessel takes its final shape.

6 WELDING PROCEDURE SPECIFICATION AND WELDER QUALIFICATION

For carrying out any welding on a pressure vessel, an approved welding procedure specification (WPS) is an essential requirement, and all the welders involved vessel have to be qualified as per ASME Section IX. The various aspects of the qualification are discussed at length in Chapter 17.

7 WELDING

7.1 General

Qualified welders shall carry out the welding as per approved procedures. However every weld has the potential for distortion. It cannot be avoided. However, if judicious precautions are taken, it is easily brought under acceptable limits. The author considers this an important aspect in pressure vessel manufacture. However it is often not given due stress in books related to manufacturing processes or welding. There are a very few isolated studies by individuals and publications on distortion, most of which are very general in nature. Therefore, various parameters that attribute to distortion and remedial action to minimize the same are also discussed in this chapter. These are based on the experience and studies conducted by the author.

7.2 Edge Preparation

The edge preparation has a very significant role in controlling distortion. When a rational design of the weld joint is made, the quantum of weld metal is be minimized, so the quantity of weld on either side of the weld can be balanced. Obviously, the associated distortion also will be less. It is generally recommended to have single-V double-welded butt joints for shells up to approximately 16 mm thickness, above which double-V double-welded butt joints are preferred. Even though the V sizes and angles are stipulated by the codes, where to apply them and where to do the first welding, etc., are not addressed. Given here are some suggestions that could drastically reduce effects due to distortion, if used judiciously:

1. For shells below 16 mm thickness, a double-welded single-V may be used. As far as possible, provide the V on the inside of the shell (see Figure 4.21A). Subse-

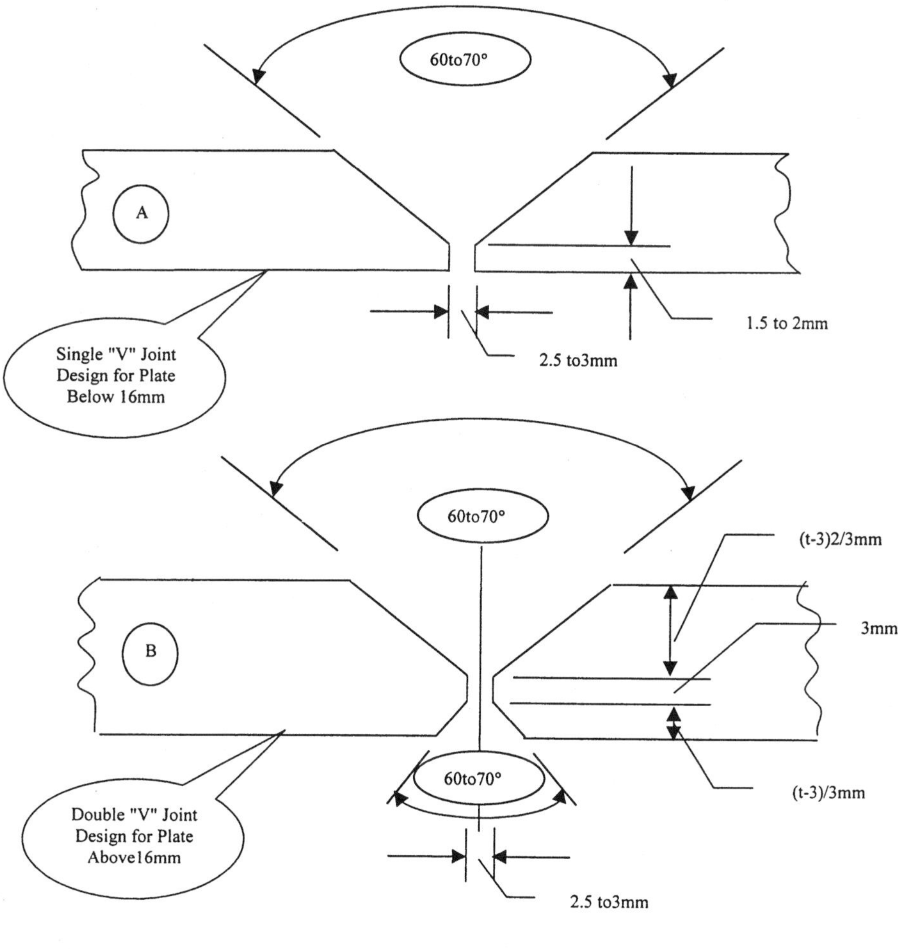

Figure 4.21 V joint preparation.

quently, the back-gouging has to be carried out from outside with a proper restraint applied to the joint from inside, as mentioned in Section 5.5. This will result in balancing the quantum of weld metal from both sides and thereby counteract the forces of contraction (see Figure 4.22A).

2. In the case of double-V welds (for thickness over 16 mm), provide unequal V, i.e., 1/3 t and 2/3 t configuration, as shown in Figure 4.21B. Provide the major V on the inside and the minor V on the outside. The welding has to be completed first from inside, and back-gouging has to be carried out from outside. Here again while doing the back-gouging care shall be taken to see that it has not gone very deep to reach sound metal. The ideal maximum is up to half the thickness of the shell, as shown in Figure 4.22B. Welding from outside shall be carried out only with restraints in place from inside, as in the previous case.

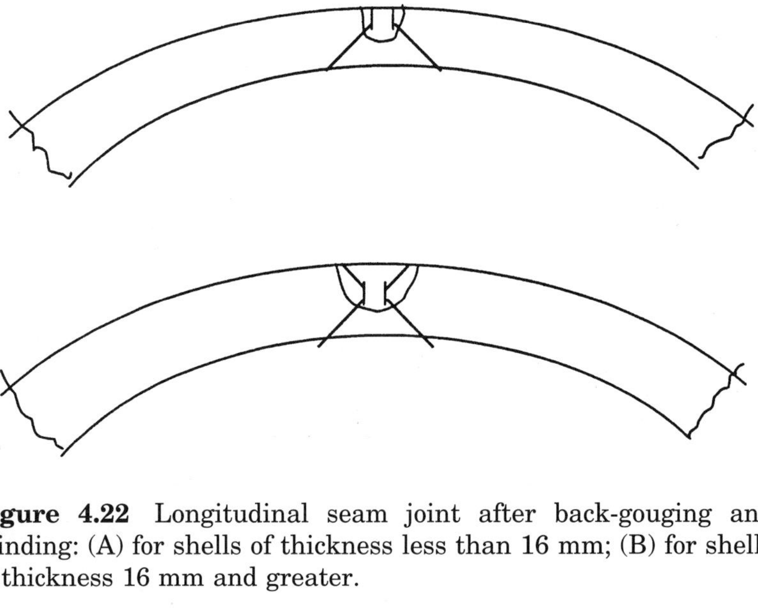

(A)

(B)

Figure 4.22 Longitudinal seam joint after back-gouging and grinding: (A) for shells of thickness less than 16 mm; (B) for shells of thickness 16 mm and greater.

3. The V angle shall be kept at the minimum possible so that the quantum of weld metal can be reduced.
4. While back-gouging and subsequent grinding, the resultant groove shall have a V angle of at least 10 to 15°, as shown in Figure 4.22, for the slag to float up to the top of the weld easily.

7.3 Tack Welding of Joints

Tack welding on the seam or the use of clamps and wedges may be resorted to for holding the joint in position. Provided a qualified welder does the tack weld, using a qualified welding procedure under the requisite conditions of preheat, it can be a part of the actual weld. In that case, its stopping and starting points shall be properly prepared by grinding prior to full welding. Tacks shall be provided at a spacing of 200 to 250 mm, with the length of tack ranging from 25 to 40 mm. The holding devices used for maintaining the shape of the shells shall be retained until the welding from the major V side is completed, as the distortion is maximum in the first few passes of the weld.

7.4 Run-In and Run-Out Plates

Two pieces of plate of 50 to 100 mm length shall be provided at the beginning and end of each seam, preferably with the same V configuration. This will help the welder to start the weld from the run-in plate, and by the time the welder reaches the shell, welding will be stabilized and the weld on the shell will be free from defects. This is possible only for longitudinal seams and has a serious bearing on the quality of the weld at the start and end of each longitudinal seam, which eventually turn out to be T joints.

7.5 Stiffeners to Maintain Circularity

All shell sections shall be provided with spiders for maintaining the circularity while rotating the shells on floor as well as on rotators. At least four spiders shall be given at both the ends of the shell avoiding the longitudinal seam area, as shown in Figure 4.23. The spiders can be of angle iron preferably with adjustment provision to maintain the circularity precisely. Apart from these spiders meant for maintaining circularity, one additional stiffener shall be provided on the weld with a preset as shown in Figure 4.24. The preset thus given will be nullified by the shrinkage forces acting against it, thereby resulting in a weld seam with peaking in or out in the acceptable range.

7.6 Precleaning of Welds

A weld seam shall be thoroughly cleaned and shall be free from rust, grease, mill scale, or any other foreign material not only on the weld but also on either side of the V groove to a width of at least 25 mm. This shall be done meticulously on both sides of the seam prior to the start of the weld on the first side. The use of a circular wire brush attached to a grinding machine could produce the desired results, whereas the simple wire brush cleaning is not sufficient as specified in many of the welding procedures.

7.7 Preheating

Based on the material of the shell section and also on the WPS, the joint may have to be preheated prior to welding. This is given in Appendix R of ASME Section VIII Div (1) (see Chapter 21, Section 15). Depending on the thickness of the plate to be welded and the temperature to be attained, suitable attachments shall be made to achieve the same. The arrangement shown in Figure 4.25A is intended for heating the long seam, and that for circumferential seam is given in Figure 4.25B. The maximum length of the attachment needed is 1000 mm, as it can be moved in the direction of the weld as the weld progresses. The heating shall be done from outside when the welding is carried out from inside and vice versa. The temperature at the joint shall be measured using temperature chalks.

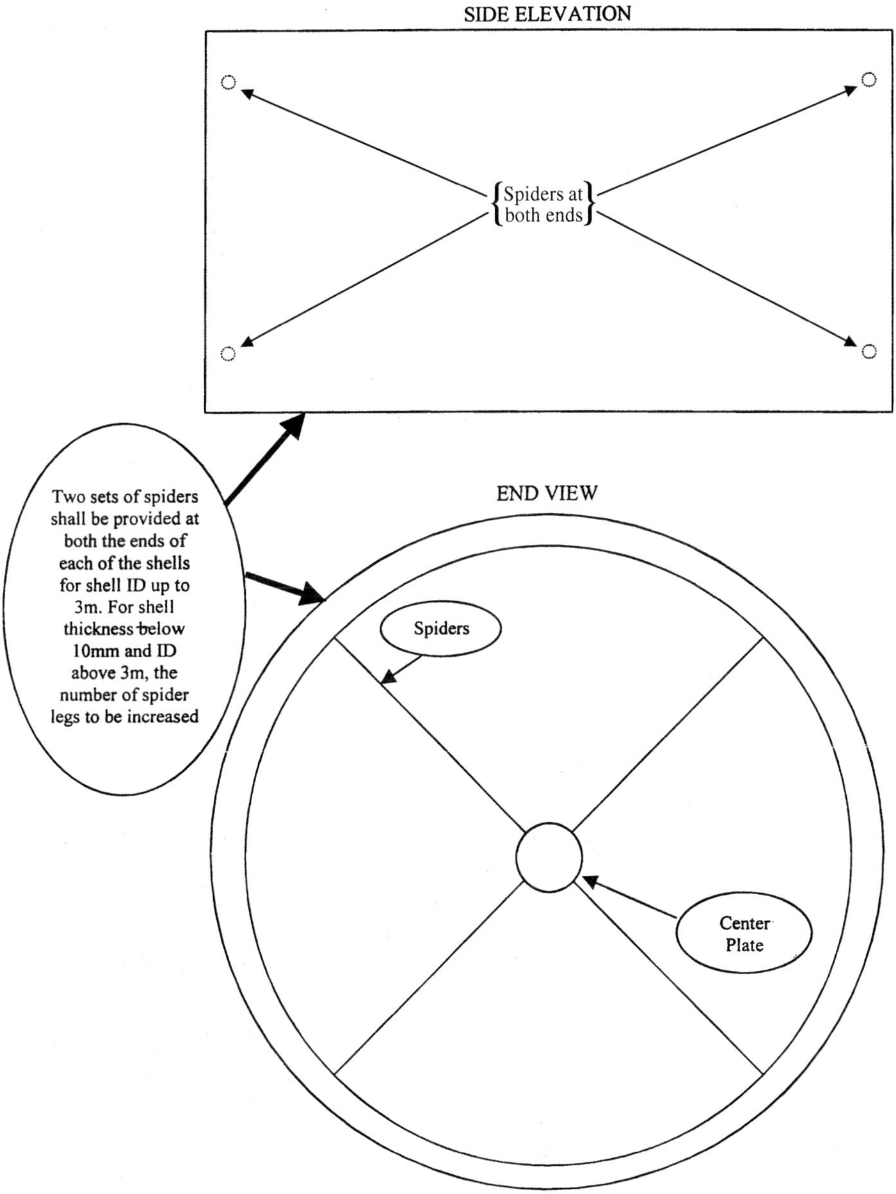

Figure 4.23 Fixtures to maintain circularity.

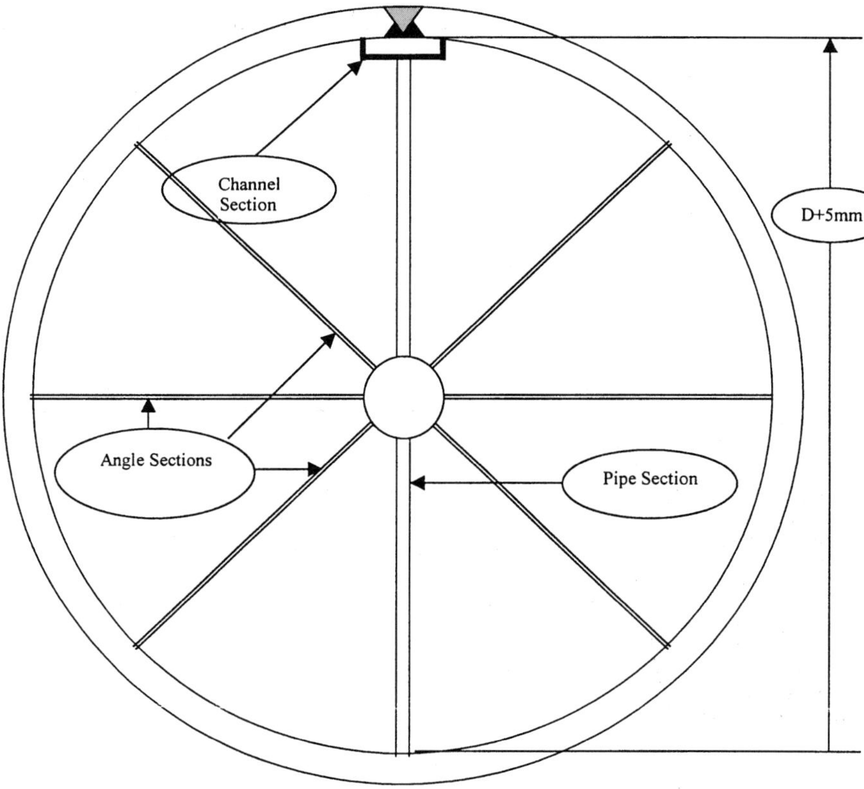

Figure 4.24 Restraint on long seam for welding from outside. *Note*: (1) The channel section shall be tack welded only on one side; (2) the pipe shall be tack welded to the channel at the top and the shell at the bottom; (3) channel and pipe of suitable size and thickness shall be used based on the thickness and diameter of the shell.

In order to rule out the possibility of measuring the skin temperature instead of the actual plate temperature, the measurement shall be done on the opposite side of the shell where the heating device is placed. In case better controls are needed, thermocouples can be provided along the weld joint at certain intervals and be monitored using recorders.

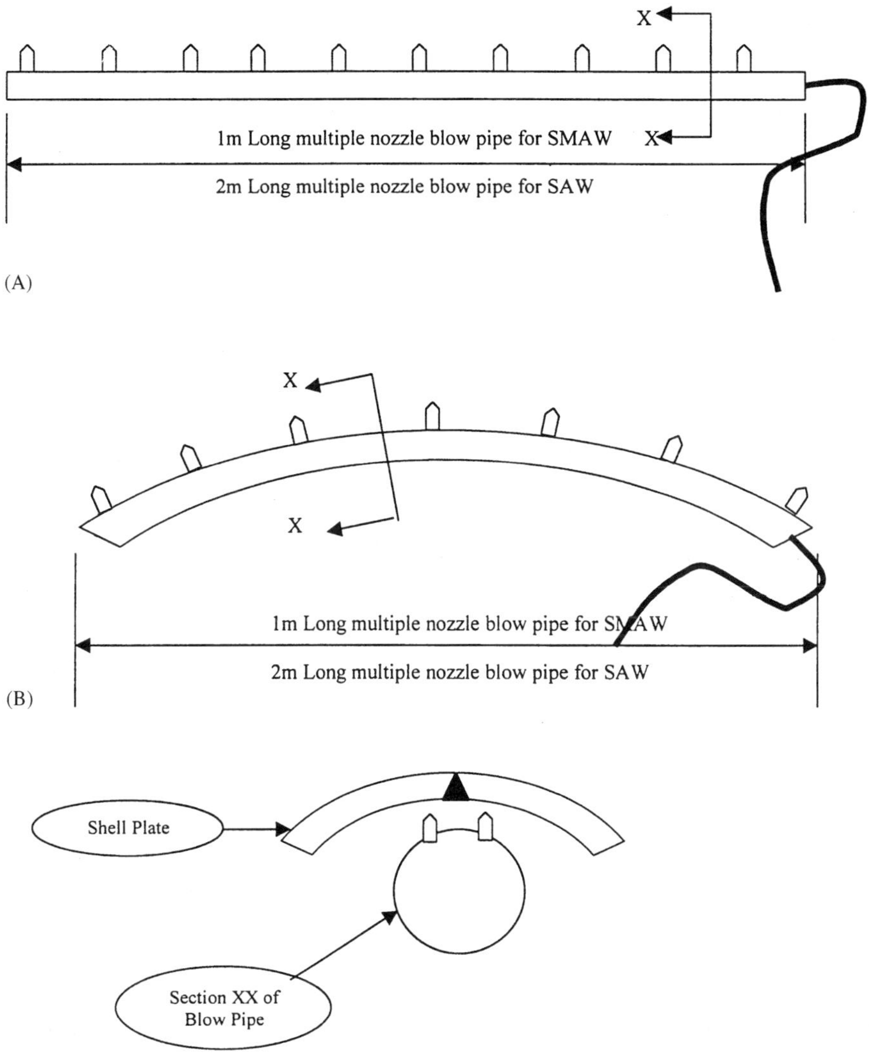

Figure 4.25 Preheating arrangement: (A) for longitudinal seams; (B) for circumferential seams.

7.8 Sequential Welding

For a longitudinal seam of approximately 1.5 to 2 m long, the weld can start from one end of the seam and finish at the other. While setting the joint for such welding, care shall be taken to see that the gap provided at the starting end of the longitudinal seam shall be exactly as per the drawing, and toward the closing end it shall be increased a little bit. For a long seam of length 1.5 to 2 m, 2 mm excess gap at the closing end can be given which will shrink to the desired level as the weld progresses toward the closing end, as shown in Figure 4.26.

When the length of the seam is more than 2 m, it is better to adopt a staggered welding sequence. For a weld seam of about 4 m, it is recommended to start welding from one end to a length of about 1 m and next from the other end for the same distance. The remaining 2 m left out at the center is welded last.

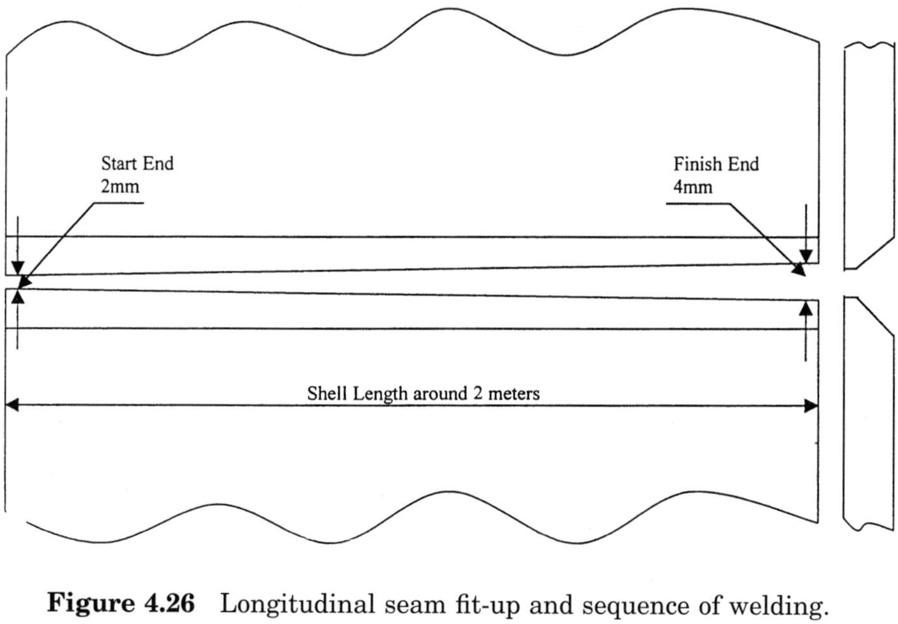

Figure 4.26 Longitudinal seam fit-up and sequence of welding.

For a thin shell where the single V is provided at the inside, upon completion of the root pass, the welded side shall be cleaned using a power brush. If the weld after the root pass is too rough, smoothing of the weld ripples may be carried out by grinding. Later the welding from inside may be completed, and outside welding may be carried out after back-gouging and providing adequate stiffeners/preset, as shown in Figure 4.27.

7.9 Interpass Cleaning

After every pass, the weld has to be cleaned using proper tools to remove all slag and spatters from the V groove. In most cases, power brush cleaning after chipping the slag is sufficient. However in the case of root run, it may call for grinding, as it is difficult to remove slag from a deep groove. When the welded surface is free from slag, spatters, and surface defects, it can be taken for depositing subsequent beads after checking the interpass temperature.

7.10 Interpass Temperature

For any welding to go on, the welded joint shall have a minimum preheat temperature and at the same time it shall not exceed the maximum interpass temperature indicated in the WPS. These two parameters can be monitored as mentioned in Section 7.7.

7.11 Back-Gouging/Grinding

In the case of longitudinal seams, it is recommended to complete the welding from inside first and then perform the back-gouging from outside. Back-gouging may be carried out either by carbon electrode gouging or grinding alone. In the case of carbon electrode gouging, it shall be ensured that all black carbon gouge marks on the gouged area are removed by grinding into sound metal by at least 1 mm. This is to prevent the

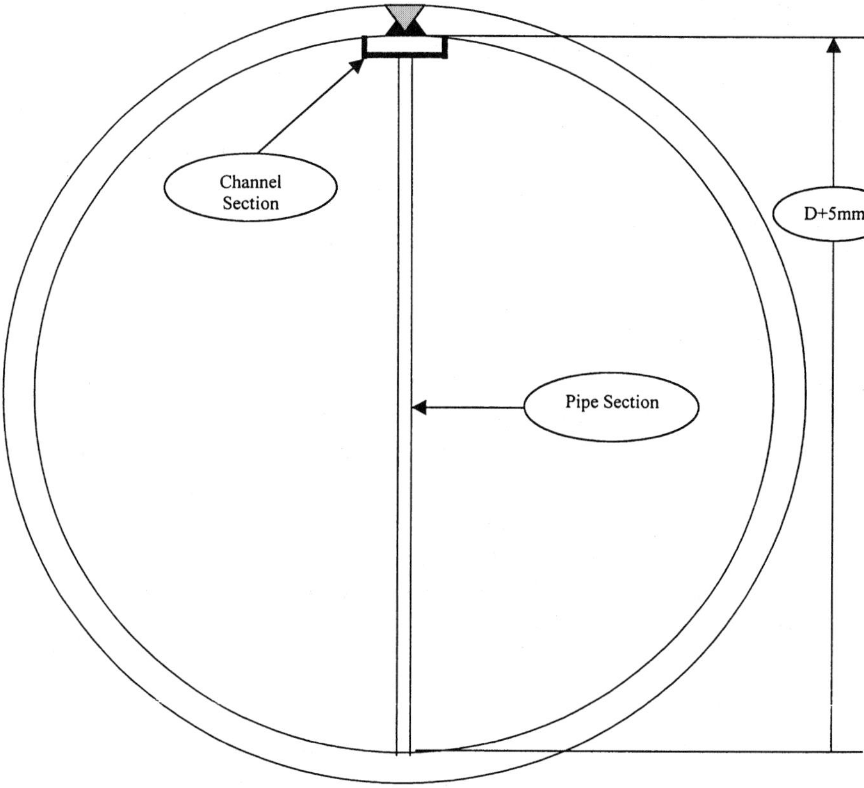

Figure 4.27 Restraint on long seam for welding from outside. *Note*: (1) The channel section shall be tack welded only on one side; (2) the pipe shall be tack welded to the channel at the top and the shell at the bottom; (3) channel and pipe of suitable size and thickness shall be used based on the thickness and diameter of the shell.

carbon particles from going into the molten metal, which may cause detrimental effects. While back-gouging, care shall be taken to see that

1. It is not going to a depth beyond half the thickness of the plate in order to reach the sound metal.

2. The width of the V is not exceeding the designed V width.

At the same time, it is important that a vertical wall for the V be avoided. The resultant V should have at least a V angle of 10 to 15° on either side, as shown in Figure 4.28, so that the slag will float up easily.

7.12 Nondestructive Testing After Back-Gouging

The LPT of the back-gouged joint is always useful in detecting gross defects like lack of fusion, cluster porosity, slag inclusions, etc., in the root run and hence is recommended for all full penetration welds.

7.13 Balancing of Welds

Care shall be taken in carrying out the back-gouging so that an unequal V joint after back-gouging from the minor V side shall never exceed half the thickness of the shell. In other words, the weld from both sides shall be almost balanced in quantum. In addition, care shall be taken to ensure that the V width is not increased beyond that specified in the design. However, the resultant V shall have a V angle of at least 10 to 15°, as shown in Figure 4.28B.

7.14 Test Coupons

UHT 81 stipulates the requirements for the use of coupon plates for verification of heat treatment in ferritic steels with properties enhanced by heat treatment. One or more test coupons representative of the material and welding may be provided for the purpose. The requirement varies with the heat treatment process. However, in most cases, the coupon plates are provided along with the long seam and welded under the same welding parameters as that of the longitudinal seam of the shell. It is expected to undergo the same type of heat treat-

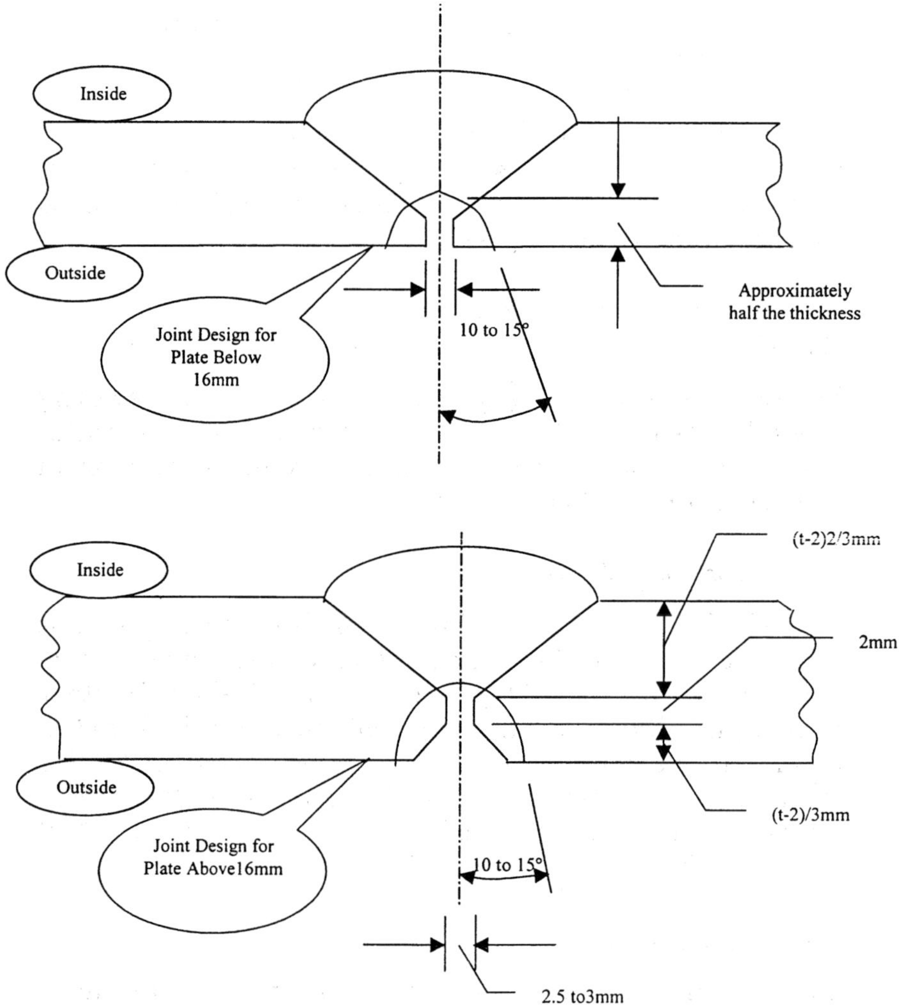

Figure 4.28 Weld groove after back-gouging and grinding.

ment later as that for the vessel and mechanically tested to prove the properties.

7.15 Final Visual Examination

On completion of welds on either side, the restraints provided during welding may be removed for ease in carrying out other jobs. However the spiders provided to maintain the circularity of shells, especially for thin shells, may be retained until the vessel is complete so that it can take such loads on its own.

The weld joints shall be inspected after the completion of welding from both sides using an inside or outside template to assess the peaking in or out of the seam. If this is within acceptable limits as per the code or client specification, visual inspection of the seam is carried out with regard to the following aspects:

1. Surface finish of the weld and its uniformity according to specifications for welded surfaces permitted in UW 35 (see Chapter 21, Section 16). However, the surface of the weld shall be sufficiently free from coarse ripples, grooves, overlaps, and abrupt ridges and valleys to permit the proper interpretation of radiographs.
2. Reinforcement of the weld shall be as per the table given in UW 35.
3. Profile of weld joint.
4. Under cuts.
5. Spatters.
6. Removal of tack metals in the nearby areas.

When all these parameters are satisfactory the weld can be released for nondestructive testing (NDT) as needed.

7.16 Nondestructive Testing of Long Seams

All seams that have undergone satisfactory visual inspection can be taken for NDT, the most prominent of which is RT. As per the applicable specification, UW11, ultrasonic examination carried out in accordance with UW 53 (see Chapter 21, Section 11) may be substituted for radiography for the final closure seam of a pressure vessel if the construction of the vessel does not permit interpretable radiographs. The technique and acceptance criteria shall be in accordance with Appendix 12 of ASME Section VIII Div (1) (see Chapter 21, Section 12). Other NDT methods include MPT and LPT, the applicable section of the code for these are UG 93 and UHT 57, respectively. For both methods, the basic requirements on methodology, evaluation, and acceptance shall be as per Appendixes 6 and 8 of ASME Section VIII Div (1), respectively (see Chapter 21, Sections 13 and 14).

The requirement of RT can be either full or spot depending on the joint efficiency considered in the design.

7.16.1 Full Radiographic Testing

The applicable sections of the code in this regard are UW 11 (see Chapter 21, Section 6) and UW 51 (see Chapter 21, Section 8). Based on the design considerations, the quantum of radiography will be decided at the initial stages of design. Full radiography is required when the joint efficiency is considered as one in design. In addition, all butt welds in shells and heads of vessels meant to carry lethal substances and all butt welds in shells whose thickness exceeds 38.1 mm shall also undergo full radiography.

However, category B and C welds of nozzles that exceed neither 10 in. size nor 1$^1/_8$ in. wall thickness do not require RT.

7.16.2 Spot Radiographic Testing

The applicable code is UW 52 (see Chapter 21, Section 9). As in full RT, spot radiography also will be decided by the designer based on the weld efficiency factor assumed for the purpose.

As per the referred code, one spot on each vessel is required for each 50-ft (15.2-m) increment of weld or fraction thereof. The minimum size of a spot shall be 6 in. (150 mm). When such a spot RT film does not meet the acceptance criteria as specified in UW 51, two additional spots in the same weld away from the initial spot shall be taken. If these two spots are acceptable and the original spot is repaired and a new radiograph also meets the acceptability norms, the joint represented by these welds can be considered as accepted. When either of the additional spots fails to meet the requirements, the entire length of weld represented by this weld shall be fully radiographed. The acceptance norms in case of both full and spot radiography are as per UW 51 for linear defects. Whereas for rounded indications, Appendix 4 (see Chapter 21, Section 10) is the acceptance standard.

8 ASSEMBLY OF SHELLS, HEADS, AND CONES

8.1 Shell to Shell

Any completed shell shall have a reference centerline on the shell, which shall be marked at least in 0°, 90°, 180°, and 270° and shall extend through the full length of the shell. Consider a shell with only one longitudinal seam whose circularity is maintained satisfactorily by providing spiders at both the open ends. Obviously the seam can be made a reference line. However, to mark the same on the shell, the following methodology shall be adopted:

1. Put a centerpunch on the center of the weld at one end of the shell.
2. Stretch a measuring tape from there and measure the outside circumference of the shell at that end.
3. Divide by 4 to get one-quarter measurement, and mark it along the circumference to obtain 90°, 180°, and 270° at the same end. The difference in dimen-

sion of each quarter shall be limited to 2 to 3 mm maximum, depending on the ID of the vessel.

4. Transfer the centerpunch put on the center of the weld at one end of the shell to the other end by water level. For this the shell shall be first kept on leveled rotators. The rotators are to be leveled by adjusting the rollers at both ends separately.
5. Make the same divisions at this end also.
6. Connect these points to obtain the four centerlines, which will be the four major longitudinal axes of the shell.
7. The same methodology has to be repeated for all the shells.

When the shells are to be assembled, the staggering of the weld joints shall be effected in the cutting plan. From the four centerlines marked on each shell, one can easily locate the staggering required, and shells can be assembled accordingly.

Since the first shell is made stationary, the second is rotated to get an offset of X mm while aligning the shells. The same process is repeated for other shells also. Upon completion of the fit-up, the following aspects shall be checked:

1. The offset or mismatch between shells all around the circumference is measured by means of a straight edge and a taper gauge. The allowable alignment tolerance after welding is as per table UW 33 (see Chapter 21, Section 5). Any offset within the allowable tolerance shall be faired at 3-to-1 taper over the width of the finished weld or, if necessary, by adding additional weld metal beyond that which would otherwise be the edge of the weld metal. However, as a rule of thumb, the allowable tolerance can be easily taken as the lowest of either 10% of the thickness of the shell or 2 mm. If shells of different thickness are joined, the thickness considered shall be that of the thicker

plate. In the case of mismatching shells within the allowable tolerance, the root face also can have the same mismatch that is to be dressed off prior to welding to facilitate positive penetration for the weld.

2. The V joint configuration is checked for uniformity, bevel angle, root face, etc., using a welding gauge.
3. Straightness of the assembled shells is checked using a stretched piano wire through the four centerlines. The possible bend that can arise in the fit-up is called the skew, or banana bow. If the straightness is beyond the limits specified in the client specification, the joint has to be rectified.

8.2 Shell to Head

The axis on the shell is marked as specified in Section 8.1. In order to align the head to shell, the centerpoint and the four circumferential points on the head have to be marked. This is done as follows:

1. At the straight face of the head, take the outside circumference and locate the four centerpoints by dividing the circumference by 4.
2. Put the head upside-down on a thick leveled plate and place two trisquares at opposite centerpoints marked. Stretch a chalked thread on top just gracing the topmost point on the head to put an impression of the line by striking the chalked thread. Repeat the same at 90° to the first line, which will result in a point that is the center of the head (see Figure 4.29).
3. Connect the four centerpoints representing 0°, 90°, 180°, and 270° to the center of the head to get the four major axes connecting to the center of the head.
4. By the same method, any angle on the head can be located for the placement of nozzles or other attachments.

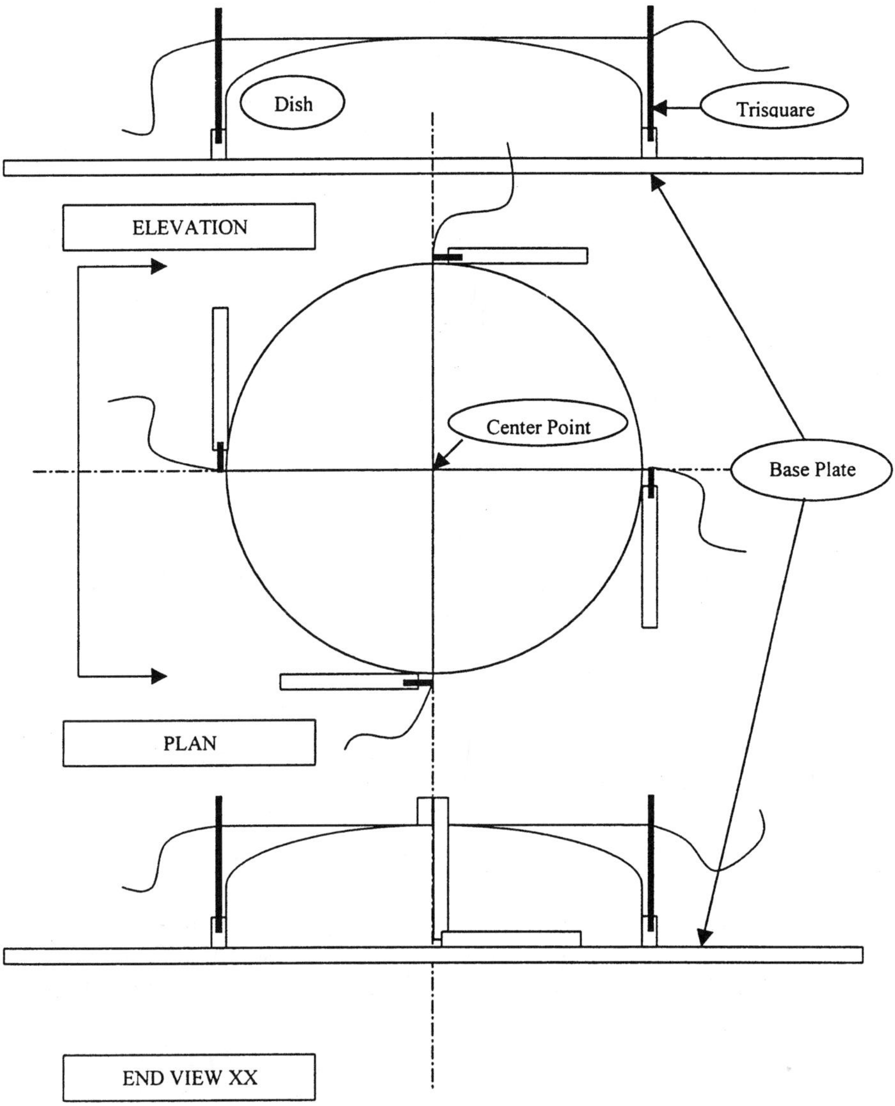

Figure 4.29 Marking of center of dish.

Alignment of head to shell is simple once the centerlines and the centerpoint are determined. Align the head and the shell by matching the centerlines. The mismatch of surfaces, V configuration, etc., shall be checked prior to welding as explained in Section 8.1.

8.3 Shell to Cone

In order to mark the centerlines of a concentric cone, the easiest way is to consider the centerline of one of the longitudinal seams as a reference line. With reference to this, take the outside circumference at both the big and small ends and divide it by 4 to get one-quarter of it and mark the other three points on the circumference. Connect the respective points at both ends to get the other three centerlines apart from the reference line, which is the fourth center line (see Figure 4.30). With respect to these centerlines, alignment of cones to the shell can be done as desired. The joint fit-up has to be checked later as in the case of shell-to-shell joints.

8.4 Fixtures and Restraints

Although all the previously described joints are circumferential joints—due to the circular shape of the section—the most vulnerable one to distortion effects is the shell-to-shell joint compared to the shell-to-head or shell-to-cone. Therefore, restraints for only the former joints are considered. However, in the case of thin shell vessels, the very same type of restraints could be extended to joints between shells and heads or shells and cones.

In order to reduce the effects of distortion, as far as possible, the major V shall be provided inside the shell, as in the case of longitudinal joints. However, in the case of closing seams, this may have to be provided on the outside due to safety reasons. The restraints that are to be provided, in both cases, are exactly the same. It is not necessary to provide any restraints on the outside while welding from inside. Whereas

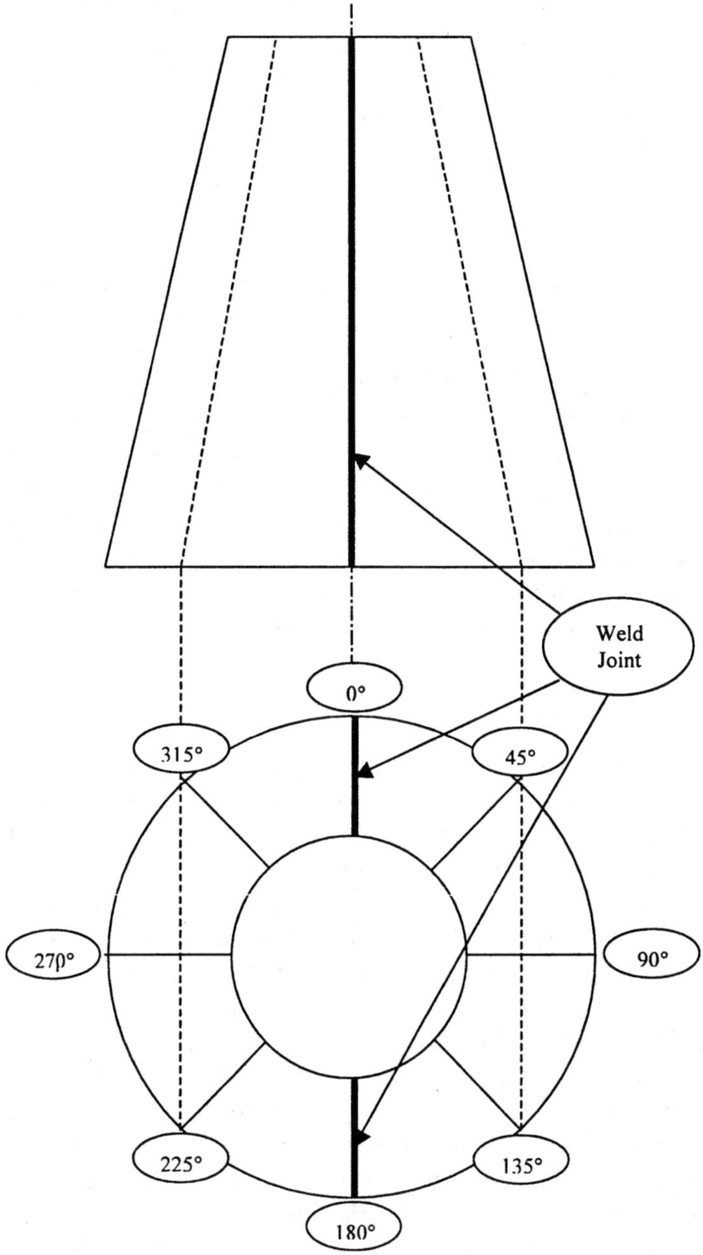

Figure 4.30 Marking of center line of cones.

during back-gouging and welding from outside, two sets of four spiders each shall be provided on either side of the joint at 40 to 50 mm distance from the joint. The number of spiders has to be increased depending on the diameter of the vessel and the thickness of the shell. The ideal arrangement for providing a controlled restraint occurs when the spiders are provided with jacking facility. The spiders can be removed upon completion and cooling down of the weld to room temperature. The details of the spiders proposed for circumferential seam welding are shown in Figure 4.31.

8.5 Welding

Most of the points mentioned in Section 7 are applicable to circumferential seams also, especially those pertaining to weld joint configuration. Apart from that, the following aspects also may be considered in the case of circumferential seams.

For circumferential seam welds, the length of weld is very large. In that case, if the welding is progressively done from one end, it may lead to a banana bow, which could be very critical in the case of tall columns. In order to reduce this, the following sequential welding methodology is suggested.

For a shell of approximately 2000 mm ID, the circumference to be welded is approximately 6280 mm. This is divided into eight segments of approximately 800 mm each along the circumference. For the root and hot passes, the segments are welded in the sequence shown in Figure 4.32. When the diameter of the shell increases, either the length of the segment is increased or the number of segments is increased or a combination of both is judiciously selected.

8.6 Nondestructive Testing

The NDT of circumferential seams is the same as that of longitudinal seams and the applicable clauses of the code is also the same.

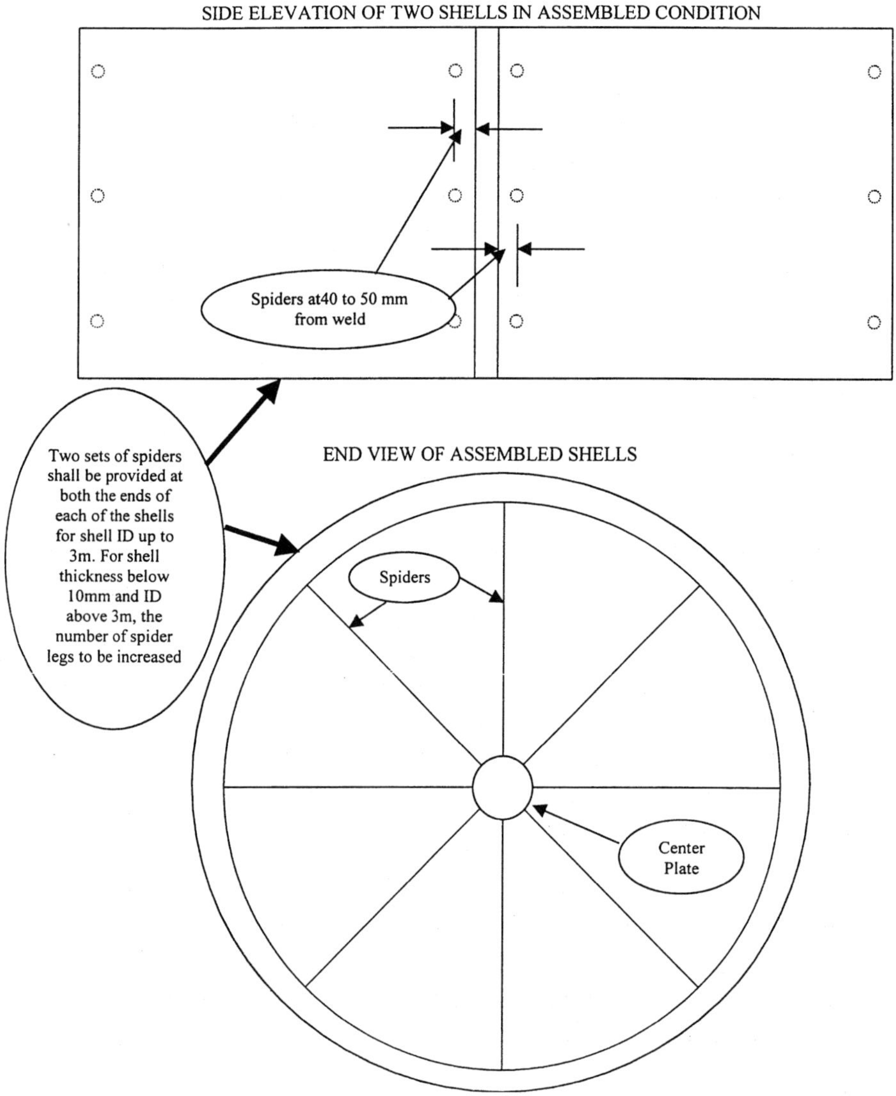

Figure 4.31 Fixtures to control distortion in circumferential seam welding.

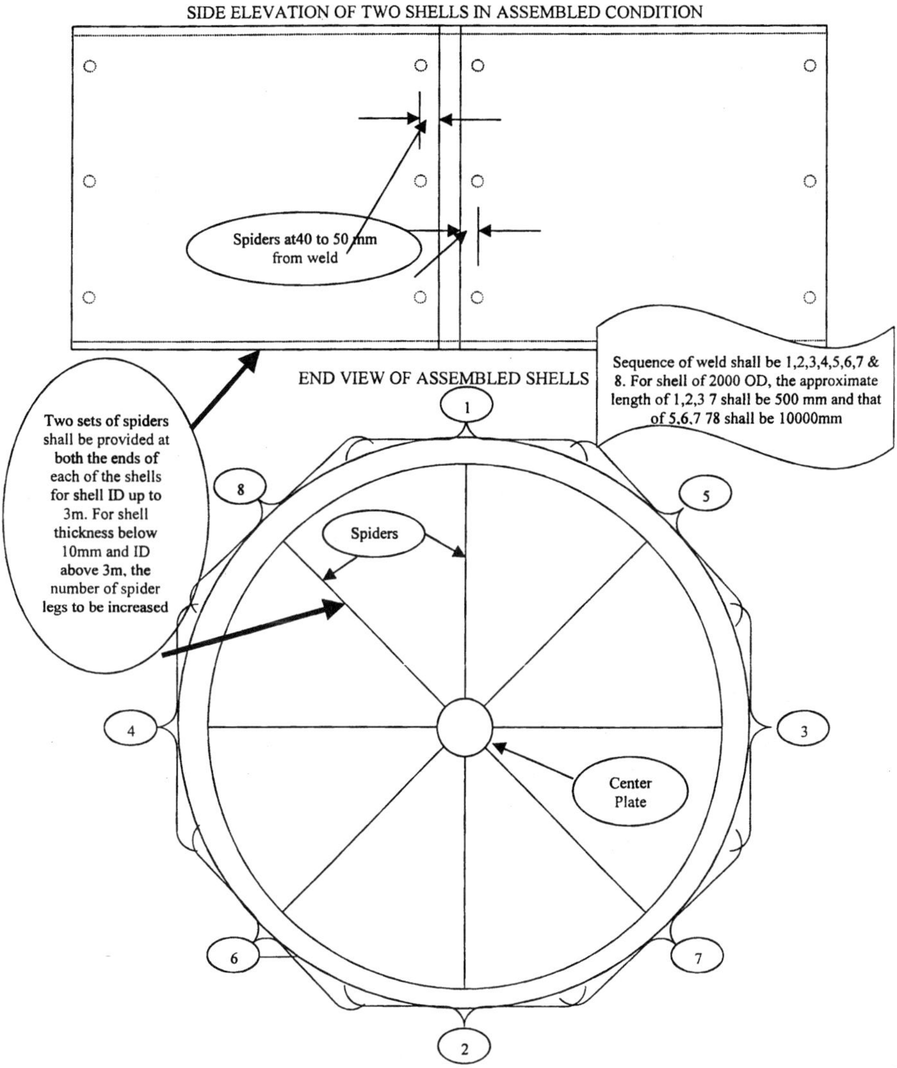

Figure 4.32 Welding sequence for circumferential seam welding.

9 MANUFACTURE OF SUBASSEMBLIES

9.1 Nozzles

Usually a nozzle consists of a self-reinforced forged nozzle neck or a seamless pipe piece welded to a weld neck flange, as shown in Figures 4.33 and 4.34, respectively. In both cases, the work involved is the same and is done as follows. The preparation of the joint is the same. The pipe or the forged nozzle is cut to the required size and aligned as per the joint design by means of external C clamps. In the case of smaller nozzles (below 100 NB), this welding shall be preferably be carried out by gas tungsten arc welding (GTAW) with compatible filler and approved welding procedures, at least for the root run and hot pass. All subsequent passes may be deposited by shielded metal are welding (SMAW) process. When the root pass is welded using SMAW, then suitable deep penetration electrodes may be used so that full penetration can be ensured.

For nozzles above 150 NB, it may be possible to do backgouging or grinding as well as welding from inside. In such cases, even for the root pass, SMAW with normal electrodes can be adopted.

In cases of pipe nozzles of size 50 NB and above, reinforcement pads of suitable thickness have to be provided as per the design. Details pertaining to the fitting of the same onto the vessel are discussed in the coming sections.

9.2 Manholes

Manholes are nothing but big nozzles of sizes 500 NB and above. Usually the pipe section used for manholes shall be a fabricated pipe from plate. Therefore, for vessels with joint efficiency one, the long seam weld is category A butt, and hence calls for 100% radiography prior to the fit-up of the circumferential joint between the flange and neck. For manholes with fabricated neck, the reinforcement pad is essential and shall be provided as per the design. In case there is any practical difficulty in placement of the reinforcement pad prior to the

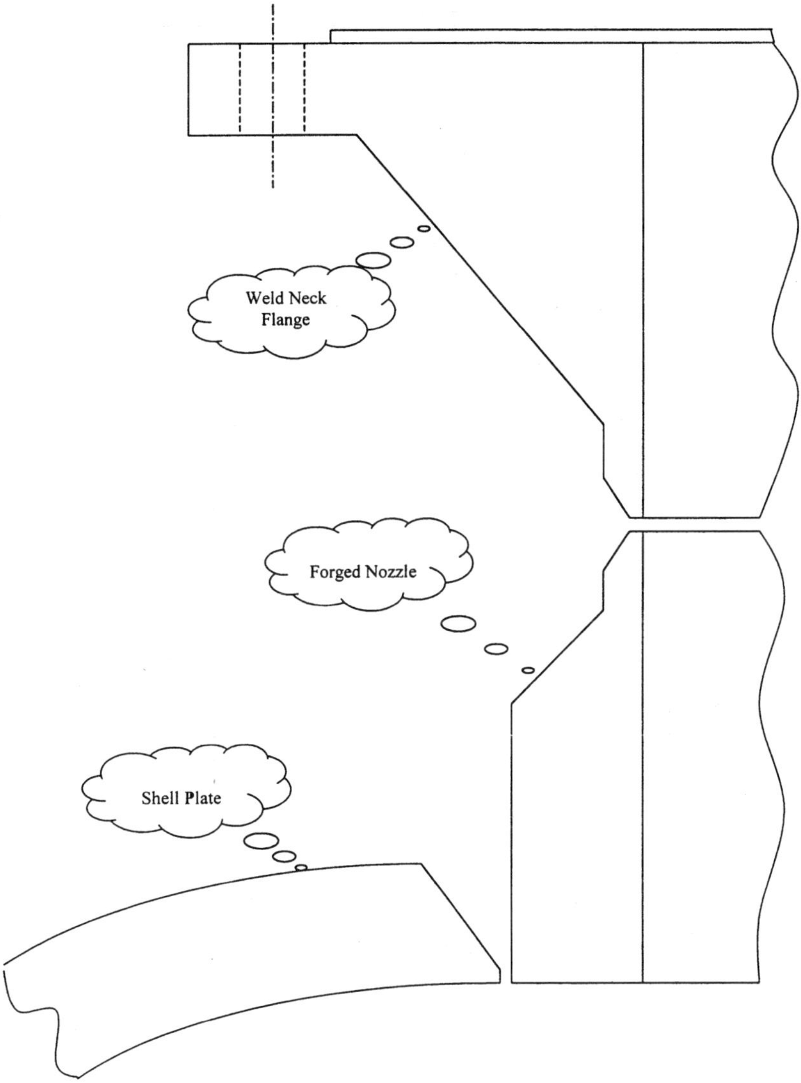

Figure 4.33 Joint detail of flange to forged neck.

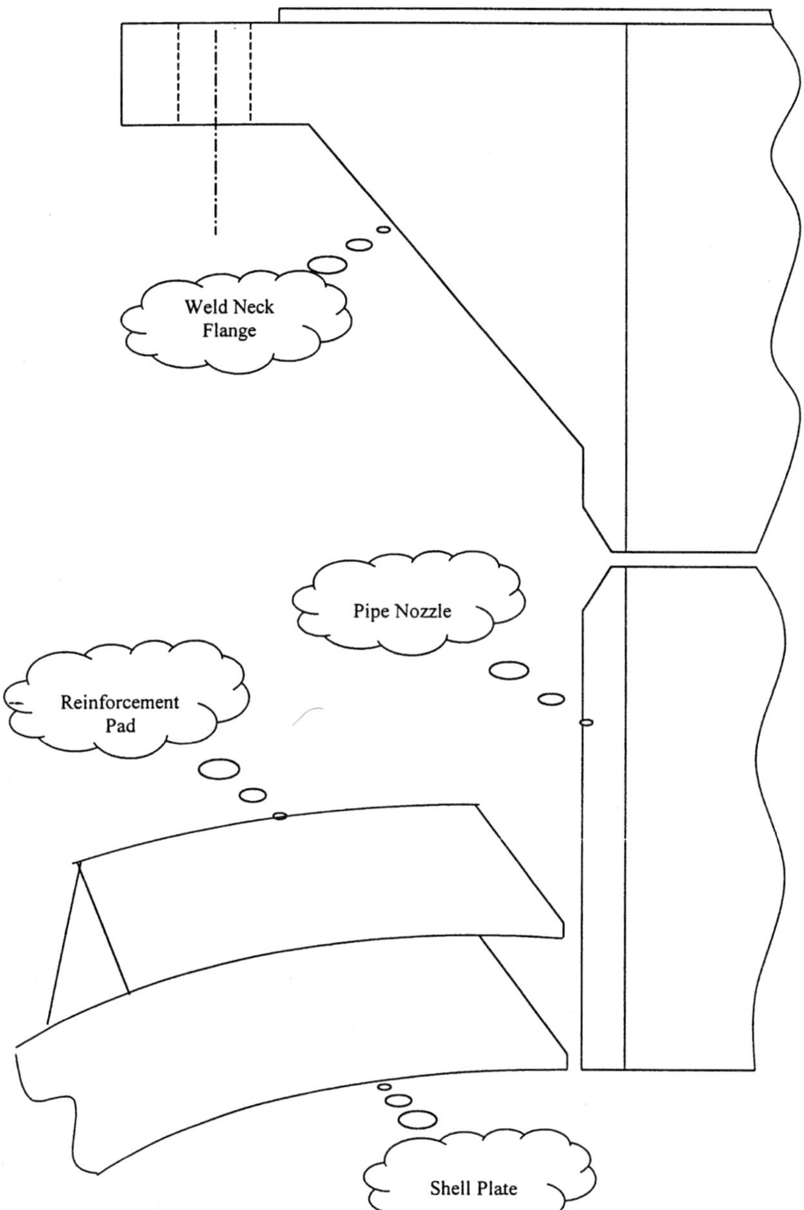

Figure 4.34 Joint detail of flange to pipe neck.

welding of manhole to the shell, it can be inserted in two halves provided there are at least two tapped telltale holes on both halves of the pad. Apart from these the manufacture of manhole is the same as that of a nozzle.

9.3 Others

The other possible fittings on a pressure vessel are saddle skirt, legs, lifting lugs, trunnions, platform/ladder clips, etc. These are directly welded to the shell if the material of the attachment is compatible to the vessel material and the design permits. If not, they are to be welded on a pad plate that is compatible to the parent plate.

All pad plates given for this purpose shall be rounded off at the corners to a radius of 25 mm radius, and the sharp edge shall be dressed to a radius of 2.5 mm minimum, as shown in Figure 4.35. The pad shall be provided with at least one telltale hole. As no opening is made on the shell, air test is not required for these pads and hence the hole provided on the pad need not be tapped.

10 FITTING OF SUBASSEMBLIES

10.1 Marking of Center of the Attachment and Cutting Line

The marking of all the attachments like nozzles, manholes, and other fittings are preferably to be done simultaneously so that any fouling of these components with existing weld seams or among them could be checked. This is checked with respect to orientation plan/elevation for vertical vessels and end view/elevation for horizontal vessels.

Care shall be taken to see that the left-hand/right-hand sides for horizontal and top/bottom ends for vertical vessels are identified. Otherwise the openings made might be on the wrong side. On satisfactory marking of the center points of these attachments, the cutting line on the shell will be marked

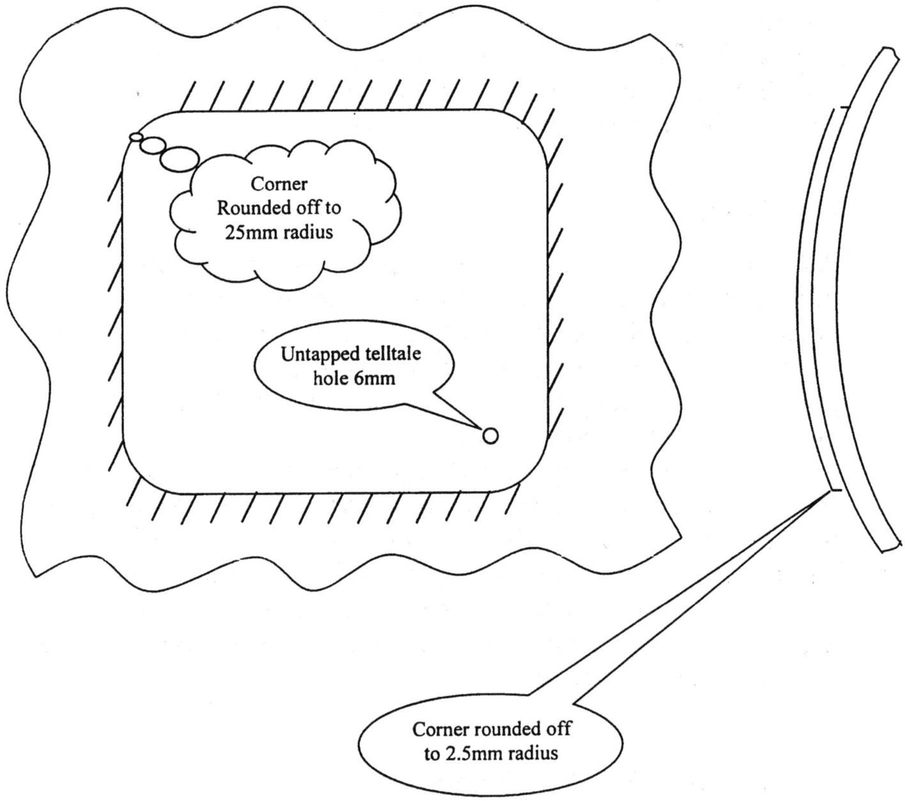

Figure 4.35 Pad plates for other attachments.

as needed, and these also have to be checked prior to the opening being made.

Consider the case of a nozzle N1 provided at 35°, as shown in Figure 4.36. Measure the outside circumference of the shell approximately at the location where the nozzle N1 is to be fitted. This corresponds to 360°. Proportionately, the outside circumference corresponding to 35° is calculated as C × 35/360. Measure out this from the reference centerline at 0° and mark a parallel line at 35°. The location of the nozzle can be anywhere on this line. Mark out the distance of 1500 mm, as

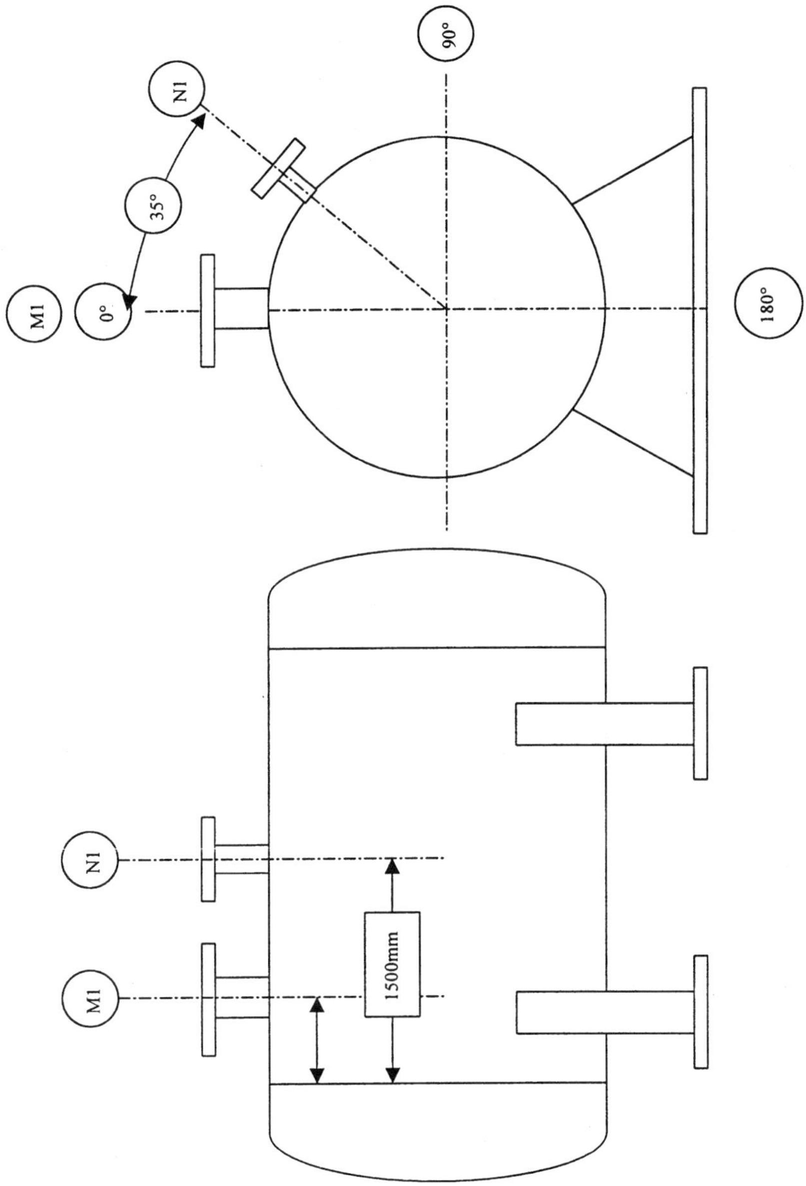

Figure 4.36 Nozzle orientation marking.

shown in Figure 4.36, from the reference tangent line to locate the centerpoint of the nozzle. Later the cutting line of the opening is also marked taking into consideration the curvature of the shell and that of the nozzle. The in-house QC personnel shall countercheck both the centerpoint as well as the cutting line to rule out possible errors in marking.

10.2 Cutting, Edge Preparation, and Grinding

The cutting is usually carried out by oxyacetylene flame in carbon steels and by plasma-arc in stainless steels. Since complicated contours are often encountered in cutting openings for nozzles, it is mostly done manually, and the uniformity of the cut edge is obtained by grinding. While grinding, the right edge preparation shall be obtained as per drawings, and uniformity in root gap, root face, bevel angle, etc., shall be obtained. In order to do the fine grinding of the hole opening, bevel angle, etc., templates for the same, as shown in Figures 4.37 and 4.38, may be used so that repair work can be minimized.

10.3 Fitting of Subassemblies

10.3.1 Nozzles and Manholes

All nozzles of 50 NB and above made out of pipe shall be provided with a reinforcement pad. If pads are to be made in one piece, then they have to be inserted to the nozzle/manhole prior to the fit-up. If the stand-off of the nozzle prevents the use of a single pad, the pad may be inserted after the fit-up of the nozzle in two pieces. The joining of the pad shall be done on the vessel with the vessel as the backing strip. In such cases, there shall be two telltale holes on both segments of the pad and shall be tapped to carry out the air test.

Prior to the fit-up of the nozzle onto the vessel, the vessel shall be leveled by adjusting the rotators, especially the idle ones. All nozzles and manholes shall be fitted as off-center to the main centerline of the vessel unless specified otherwise.

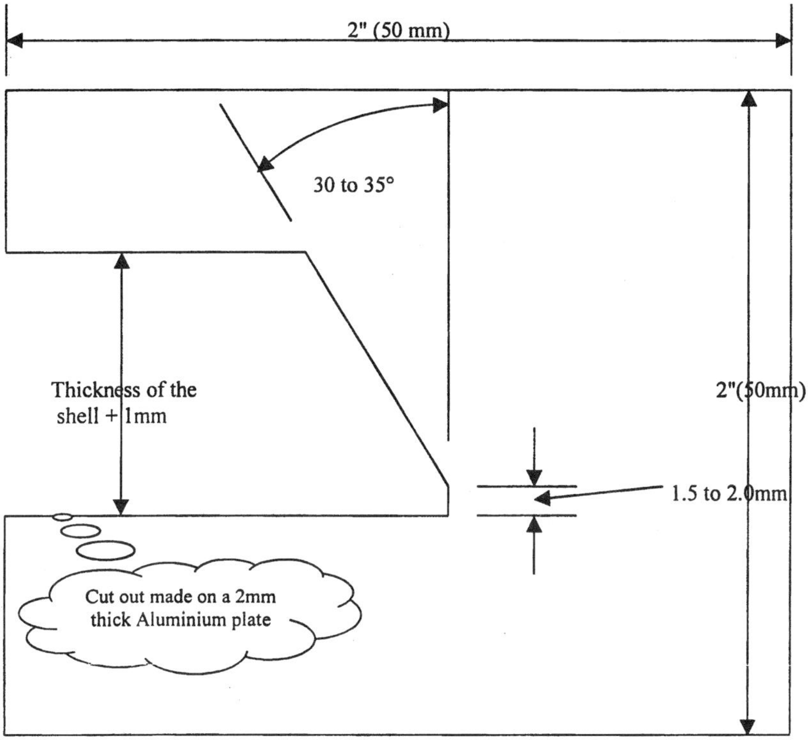

Figure 4.37 Templates for edge preparation of openings on shell and head (for single-V preparation).

For the fit-up of nozzles/manholes, the following steps shall be followed:

1. The position where the nozzle is to be fitted is placed either at the top or on the sides depending on the size of the vessel and that of the item to be fitted. The vessel is leveled again in that position in two directions (along the principal axis of the vessel and perpendicular to it).
2. Fit up the nozzle with the required stand-off and the flange holes off-center to the main axis of the vessel.

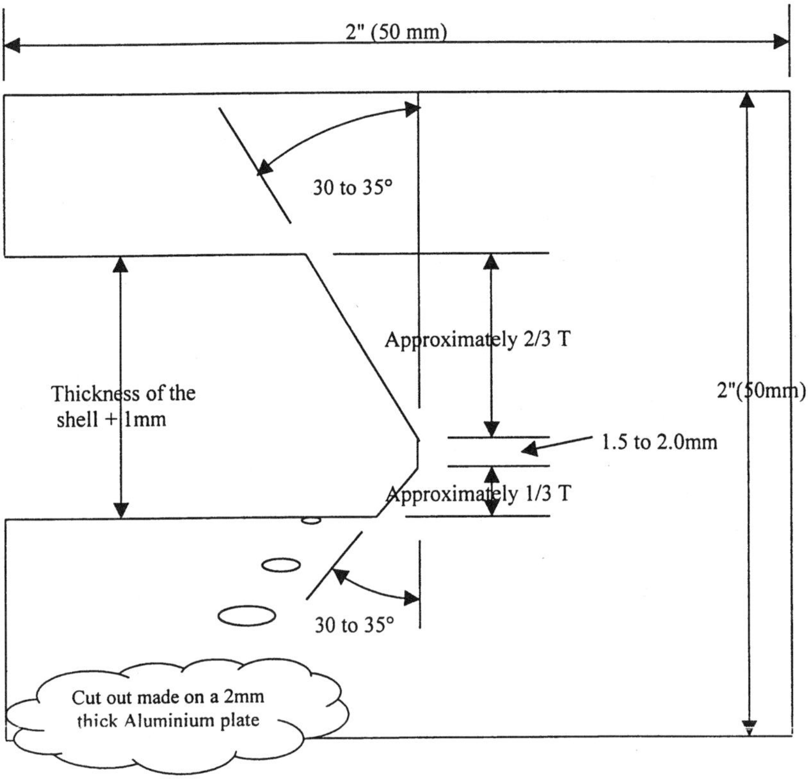

Figure 4.38 Templates for edge preparation of openings on shell and head (for double-V preparation).

3. Inspect the level on top of the flange in both the directions and if it is satisfactory (with in tolerance specified by code and client specifications), tacks are given either at the joint or by means of external cleats to hold the nozzle in position. In both the cases, the tacks shall be strong enough to hold the nozzle in position during welding.

10.3.2 Others

The items considered here are pad plates provided for the placement of attachments like saddles, legs, ladders, platforms, pipe supports, etc., to the vessel. As openings are not made in these cases, the fitting and welding are very simple compared to nozzles and manholes. However, a few good engineering practices are to be followed:

1. As far as possible, the edges of the pad plates shall be rounded off to a radius of 25 mm minimum.
2. A telltale hole shall be provided at a corner of the pad. As no testing is envisaged for the same, it can be a straight hole of 5 or 6 mm diameter without any tapping. This will permit the free flow of air entrapped between the pad and shell during welding.
3. The gap between the pad and the shell shall be maintained at 0.5 mm maximum, as is also the case of other reinforcement pads.
4. All the sharp edges on the pad shall be dressed to a radius of 0.25 mm.
5. Provide only the designed size of the weld, as shown in the drawing.
6. As openings are not made on the shell, it is presumed that the shell has sufficient strength at the location where the pad is placed. Therefore, restraints as proposed in the welding of nozzles are not necessary. However, if the size of the attachment weld is too big compared to the shell thickness, it may be better to provide restraint during welding, as in the case of nozzle welding.

10.4 Precautions Prior to Welding

Due to the opening made in the shell for the attachment of a nozzle or manhole, the vessel will become weak in that area. Therefore, during welding of the nozzle/manhole, distortion will be severe, and sufficient precautions have to be taken during welding. The distortion occurring due the welding of a nozzle or manhole to the shell is called cave-in of nozzles, and the precautions will only help to bring it down to acceptable limits. In almost all the weld joint designs, due to safety considerations, maximum welding will be done from outside. Due to this reason, the cave-in will be significant, and to minimize the same, the only effective way is to provide restraints as follows at fit-up itself so that the nozzle after welding shall be within acceptable limits on stand-off, tilt, and cave-in.

1. During fit-up always provide 3 to 5 mm extra stand-off depending on the nozzle size.
2. In case of nozzles above 300 NB and for manholes, the same can be 8 to 10 mm.
3. Provide strong gusset support all around, as shown in Figure 4.39.
4. Provide internal support as in Figure 4.39 to minimize cave-in during welding of nozzles from outside.
5. The support shall be maintained until the full welding of the nozzle is complete.
6. Adopt sequential and staggered welding to minimize uneven tilt of the nozzle.

10.5 Welding of Subassemblies

While welding subassemblies like nozzles and manholes it is always preferable to adopt staggered welding to minimize tilt and cave-in of the nozzle. The length of weld to be done at a spot is decided based on the size of the nozzle. For a nozzle of size 200 NB, the approximate circumference shall be 600 mm, and hence one segment can be about 50 mm for segments 1 to 4 and 100 mm for segments 4 to 8 (see Figure 4.40 for details). The sequence shall be meticulously followed for the root run and the hot pass. It would be desirable to continue this staggered welding technique at least for the complete weld from outside. Afterward, upon completion the weld up to the parent metal level, the weld is to be cleaned and dressed, if needed, to facilitate liquid penetrant (LPT) or magnetic particle testing (MPT).

On satisfactory completion of LPT or MPT or both, the pad is to be seated and inspected for any gap all around. Nec-

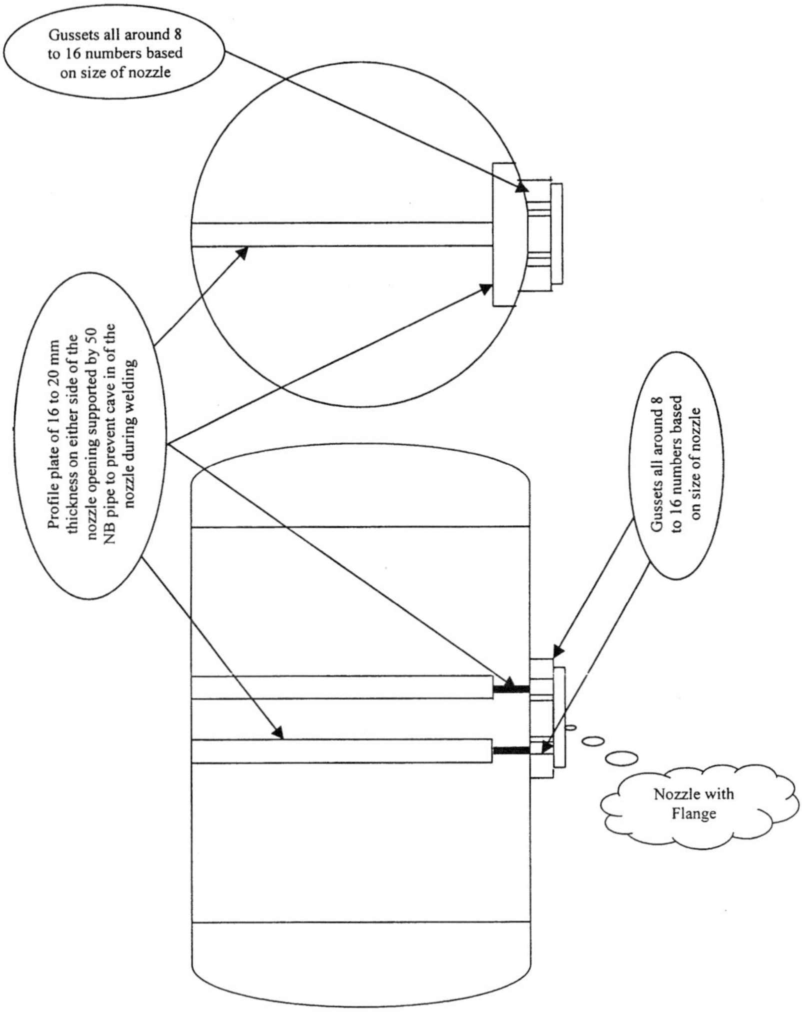

Figure 4.39 Restraints and gusset supports for nozzles during fit-up and welding. *Note*: Stand off of the nozzle at the time of set up shall be maintained as required plus 5 mm for nozzles below 8″ (200 NB) and for higher at +10 mm.

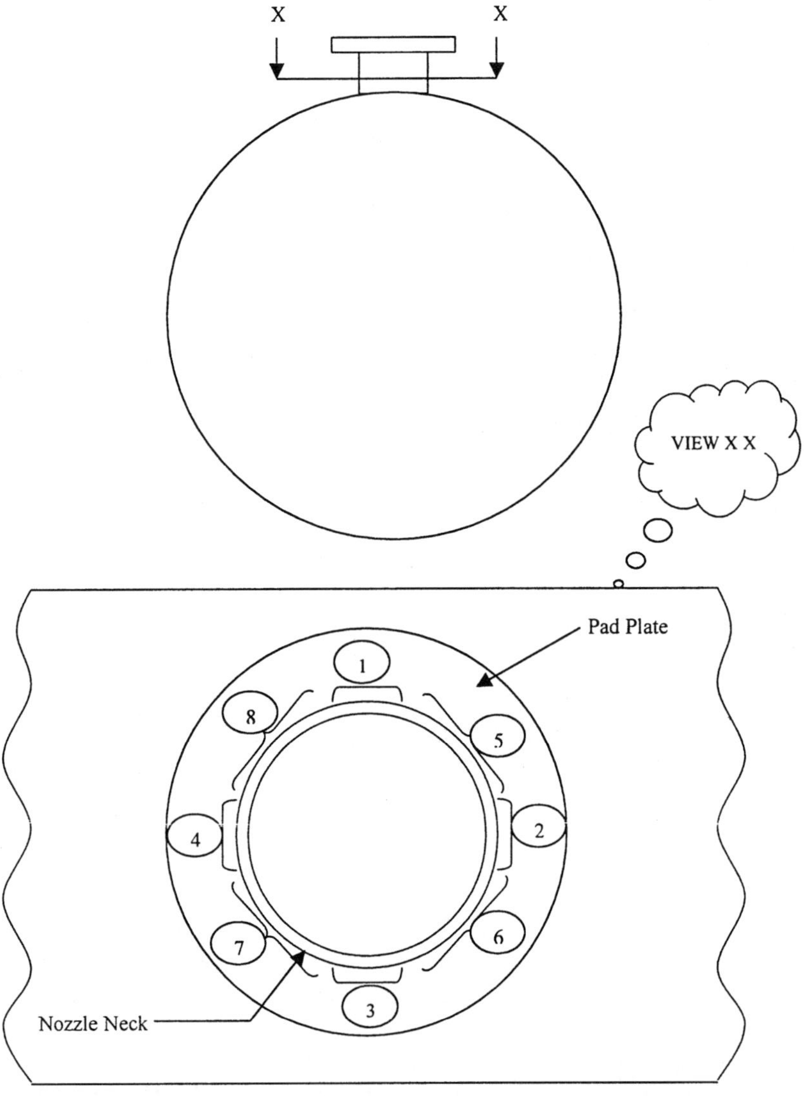

Figure 4.40 Sequence of welding for nozzle to shell/pad welding. *Note*: Sequence of welding shall be 1,3,2,4 and 5,7,6,8 for all the passes. 1,2,3,4 are smaller segments and 5,6,7,8 are larger segments. For pad to shell welding also, the same sequence may be followed for the first pass.

essary grinding shall be done to get the seating with a maximum gap of 0.5 mm.

Once the pad is fitted up, the sequential welding can be followed there also. After completing the weld of both the pad to neck and pad to shell, these joints also shall be subjected to LPT as an additional precaution.

After clearing the back-gouged weld joint by LPT, it may be filled up to the required level and the corner shall be rounded off, as shown in Figure 4.41. After dressing the weld smooth with the shell, the profile at this location shall be inspected with a standard template. The surface is to be inspected for surface defects visually as well as by LPT or MPT or both as applicable.

10.6 Qualification Requirements

If intricate welding in difficult positions is to be carried out, even though a 6G qualified welder can technically carry out the welding, depending on the constraints prevailing at site, a specific qualification can be asked for the welder intended for the job. This shall simulate all the restraints and positions involved in the actual job.

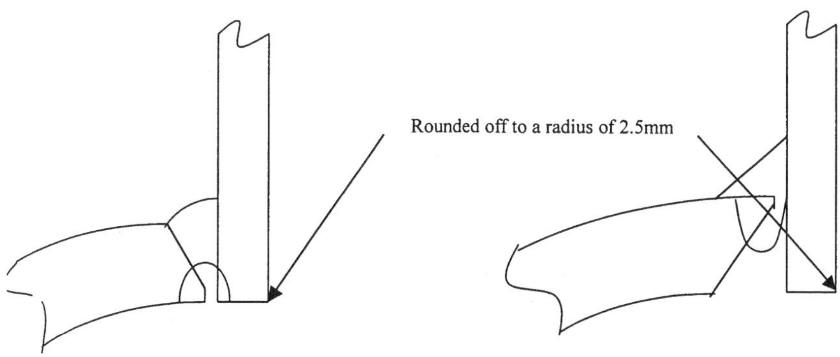

Figure 4.41 Welding and finishing of nozzle to shell welding from inside.

For example, when the nozzle stand-off is less for welding the nozzles with wide flanges, a restricted welder qualification test, as shown in Figure 4.42, may be given. The welders may be asked to do the root pass and if sufficient penetration is not produced, they are disqualified. Since making good penetration under these conditions is the test, it is pointless to carry on further. When it is essential to have a test record, the test may be completed and a radiograph of the specimen may be kept as a record.

When other types of restraints are arising in real life, such conditions shall be simulated and welders are to be qualified under such restraints over and above the normal qualification as per ASME Section IX for carrying out those specific jobs.

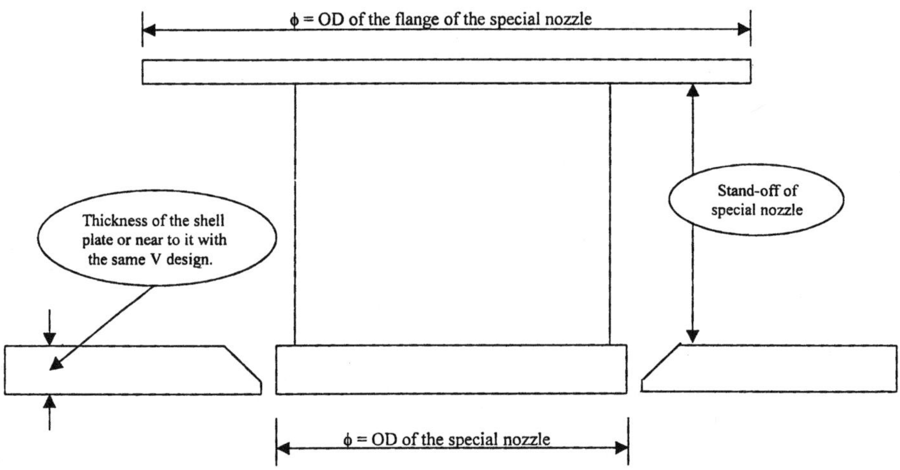

Figure 4.42 Restricted welder qualification test (WQT) specimen for nozzle to shell weld. *Note*: The weld is to be inspected for penetration upon completion of root and hot passes. In case record is needed, it may be radiographed after completing the welding from the "V" side, back-gouging and welding from the other side.

EMBLEM

DIMENSION REPORT

Order No.		Sheet	of
Purchaser		Attachment to Certificate No	
Description		Date	
Drawing No(s)			

Sl. No.	Description	Dimension (mm)	
		As Per Drawing	Actual

Figure 4.43 Dimension report.

10.7 Dimensions Record

On completion of the vessel, all salient dimensions of the vessel shall be recorded in the dimension report, whose format is enclosed as Figure 4.43. The recorded dimensions shall be verified with the general vessel tolerance furnished by the consultant and in all cases shall be within those specified. In case the dimensions recorded are beyond the tolerance stipulated, this shall be reported to the client. Their acceptance of the deviation is necessary prior to the dispatch of the vessel.

5

Nondestructive Testing

1 GENERAL

The most commonly used nondestructive testing (NDT) methods in pressure vessel manufacturing are radiographic (RT), ultrasonic (UT), magnetic particle (MPT), and liquid penetrant testing (LPT). For carrying out any NDT on a pressure vessel, a written procedure is necessary. This shall address all the salient parameters of the testing proposed during the actual performance of the test and be accompanied by the required forms used for NDT reporting. Sample procedures are discussed at length in Chapters 11 through 14.

In any type of NDT, two aspects are important:

1. The witnessing of the test as and when needed.
2. Review of reports and records of the test.

For most commonly used NDT methods except RT, testing is to be witnessed by a competent person, as there is no resulting positive evidence or record for the performance of the test. For RT, a competent person shall review the radiographs produced from the work. In all cases, reports have to be generated and signed off by all concerned parties. These reports are to be reviewed again for completeness as detailed in the coming sections.

2 REVIEW OF REPORTS FOR COMPLETENESS

A weld map indicating all the welds shall be prepared, as shown in Figure 5.1, and each seam shall be identified distinctly. By verifying the reports of each of the seams and observing remarks against each segment of the seam, the pending repairs can be quantified and repaired satisfactorily. This shall be carried out for each seam and a summary report to this effect, as shown in Figure 5.2, shall also be made for easy access to the reports. All reports shall accompany the summary in the ascending order of the report numbers.

3 RELEASE FOR POSTWELD HEAT TREATMENT

On verification of NDT reports, if it is confirmed that all the required NDT has been completed with satisfactory results, the vessel can be released for postweld heat treatment (PWHT), provided all the components that are to be directly welded to the shell have been completed. For this, each and every component in the bill of materials (BOM) of the drawing

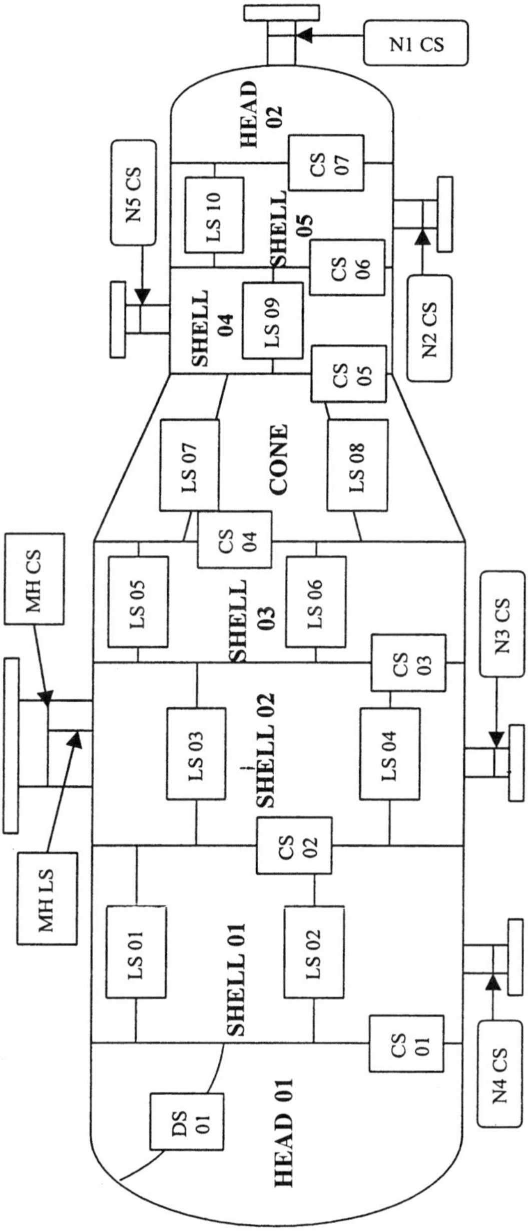

Figure 5.1 Weld map.

EMBLEM

NDT SUMMARY REPORT

Order No.		Sheet	of
Purchaser		Attachment to Certificate No	
Description		Quantity	
Drawing No(s)		Date	

Sl. No.	Seam Identification	Requirement of NDT	Details of NDT with Report Reference			
			RT	UT	MPT	LPT
1	DS1	100% RT	R.No.1,5	--	--	--
2	LS1	100				
3	LS2					
4	LS3					
5	LS4					
6	CS1					
7	CS2					
8						
9						

Note:- Certificates for Back Chip DP are not reported in this Summary. This will be covered in the general reference made on the compliance to the code of in the certificate of manufacture.

Figure 5.2 Nondestructive testing summary report.

has to be physically identified on the vessel. Loose items that are not directly welded to the vessel can be ignored for the time being. However, it is better to have all the blind flanges bolted to the vessel to protect the serrated area of the flanges from the direct impingement of the flame. Providing a temporary blind with aluminium gasket in between with at least a few bolts in position to hold the blind shall protect the other open nozzles also.

In order to avoid any possible rework due to leaks occurring in the hydraulic test, at the discretion of the manufacturer, a preliminary hydrostatic test prior to PWHT can be carried out. On many occasions, this test is carried out at a lower pressure than the actual test pressure even though the manufacturer can do it at the test pressure itself. The usual limit for this is either at the working pressure or 1.25 times the design pressure, as per the opinion of the third-party inspector.

4 NONDESTRUCTIVE TESTING AFTER POSTWELD HEAT TREATMENT

When the contract calls for carrying out the NDT after the PWHT, it shall be carried out on the same locations and with the same parameters as used earlier. The records of both the tests have to be cross-checked for any new defects or enlargement of previously accepted defects. In case any propagation of existing defects is observed, those areas have to be investigated in detail and reasons for such an occurrence are to be established. Suitable repair methodology has to be developed and approved by all concerned. After carrying out the repair, once again the proposed NDT has to be carried out. If the repair procedure calls for PWHT, once again it has undergo NDT and PWHT as stipulated in the contract.

6

Pad Air and Preliminary Hydrostatic Testing

1 PAD AIR TEST

1.1 Need

On completion of welding the nozzle to shell, pad to nozzle, and pad to shell, it is recommended to carry out an air test of the pad. This will reveal any through holes in any of the welds mentioned. Even though this test is not mandatory, it is invariably carried out to detect defects in welding which otherwise could be detected only during the hydrostatic test. In such cases, the repair of the defect becomes too expensive in terms of money and time. The test becomes very significant in the case of thin-walled vessels where the number of passes of weld is fewer. The test is carried out upon completion of all welding and the dressing of the nozzle welding to form the contour as desired in the drawings and shall obviously be before the post-weld heat treatment (PWHT).

1.2 Methodology

As mentioned earlier, all reinforcement pads shall be provided with at least one number of tapped (5-mm diameter) telltale holes for this purpose. Air is let in through a control valve at

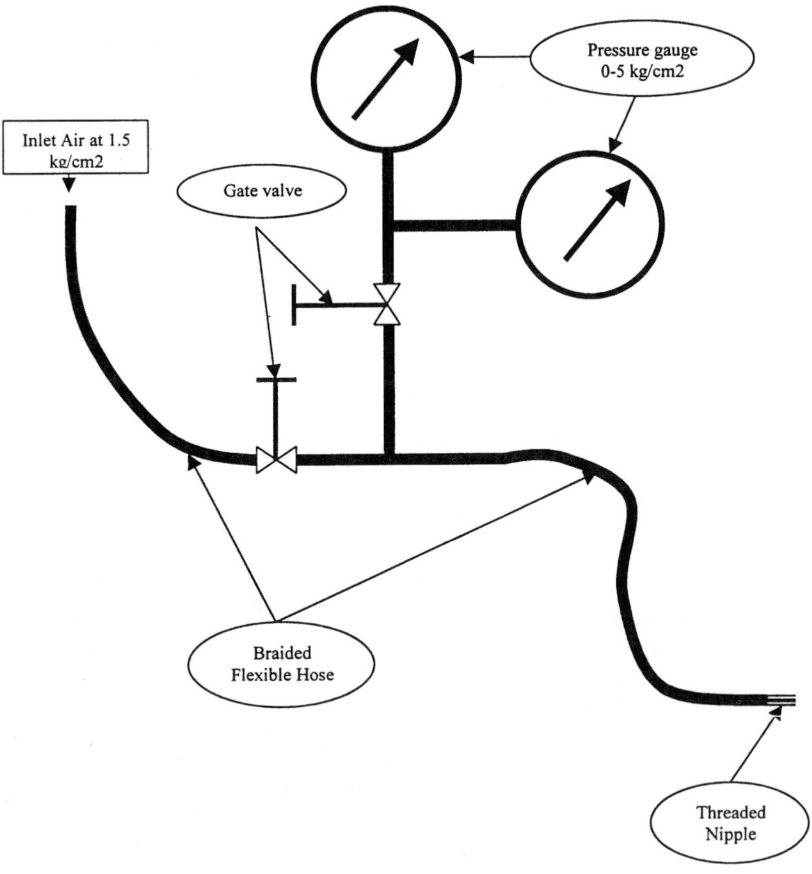

Figure 6.1 Arrangement for pad air test.

a pressure of about 1.5 kg/cm^2 (22 psi). The attachment required for the purpose is as shown in Figure 6.1. The air shall be preferably at a pressure of about 1.5 kg/cm^2 itself, or the safety valve provided shall have sufficient discharge capacity to relieve the high-pressure air from the source.

Once the air reaches 1.5 kg/cm^2 pressure, the valve is closed. If the system is leak proof, it will hold the pressure until the inspection is over. Later, a soap solution with a lot of bubbles is applied on all the welds and observed for bubbles.

In case bubbles are observed, the location is marked using colored wax crayons and later subjected to NDT to assess the magnitude of the defect. Repair procedure is worked out on the basis of the estimated magnitude of the defect. The defect is removed by grinding and subsequently filled up as mentioned in the applicable WPS. The test is carried out again after the repair.

1.3 Records

It is possible to carry out the air test of all the pads in one stretch. This is to be recorded in the format enclosed as Figure 6.2. The details of the nozzle with a brief description of the surfaces examined will serve the purpose. This record will later form a part of the MDR.

2 PRELIMINARY HYDROSTATIC TEST

2.1 Need

For vessels that are postweld heat-treated, it is recommended to have a preliminary hydrostatic test to rule out the possibility of a defect revealed during the hydrostatic test. This test is to be carried out after the satisfactory completion of NDT and pad air test, but prior to PWHT. The test pressure suggested is 1.2 times the design pressure if the final hydrostatic test pressure is at 1.3 times the design pressure. Otherwise it is very safe to carry out the test at the working pressure itself.

EMBLEM

AIR LEAK TEST REPORT

Order No.		Sheet	of
Purchaser		Attachment to Certificate No	
Description		Date	
Drawing No(s)			

Test Pressure		Holding time	
Pressure Gauge Range		Solution Used	
Pressure Gauge Calibration		Post-Cleaning	

DETAILS OF ITEMS TESTED
Remarks

Figure 6.2 Air leak test report.

2.2 Methodology

As the test is the same as that of the final hydrostatic test, the very same methodology shall be adopted for this purpose, as described in Chapter 8.

2.3 Records

The very same report shown in Chapter 8 can be used here also. However, there shall be a clear mention about the stage at which the hydrostatic test was carried out.

7

Postweld Heat Treatment

1 GENERAL

Postweld heat treatment (PWHT) of the vessel is decided based on the material of construction, fabrication methods, thickness of shell sections, etc., as contained in clause UW 40 (see Chapter 21, Section 17). Usually, vessels of 1½ in. (38 mm) and over in thickness have to undergo PWHT. Those in the range of 1¼ to 1½ in. (32 to 38 mm) need not go for PWHT, provided preheat at a minimum temperature of 200°F (93.3°C) is given during welding. Further, vessels that are subjected to low temperatures shall also undergo PWHT irrespective of thickness. As this is a design criterion, it is not elaborated here. However, as this requirement is indicated in the drawings, we try to carry out the PWHT according to the provisions of the code.

2 PROCEDURE

In any procedure for PWHT, the following aspects shall be addressed.

2.1 Method of PWHT

Depending on the size of the vessel and capacity of the furnace (size and maximum temperature), a full or partial PWHT is performed. In full PWHT, there are two types of firing methods. The most common one is the furnace PWHT wherein the vessel is loaded inside the furnace and heated to the required level in a single firing as per the previously laid down procedure. This is the most desirable type of PWHT because all parameters in the heating, soaking, and cooling cycle can be controlled well. However, the availability of such furnaces are the only constraint. Both oil firing as well as electrical heating could be used for this. If PWHT in one go is not possible due to the size of the vessel, UW 40 permits the use of part-by-part PWHT with sufficient overlap of the heated zones.

In the second method, the vessel itself is made the furnace by providing burners at appropriate points and by giving insulation all around the vessel. The method is called internal firing and is very much dependent on the skill of those that perform this feat. The method of heating adopted is generally oil firing with an additional electrical- or oil-fired heating system at anticipated cold spots.

The third method is to do the PWHT of welds alone when the design permits, using electrical resistance heating. Here again proper overlap between two PWHT zones shall be given as envisaged in UW 40.

2.2 Heating Rate

The heating rates to be followed are given in UCS 56 for carbon steel vessels and UHT 56 for steels whose properties enhanced by heat treatment. The rate specified in UCS 56 is above 800°F (427°C), which is the maximum permitted furnace temperature at the time of loading the vessel. The heating rates recommended are based on the thickness of the shells. As thickness increases, the rate falls. As per UCS 56, the rate suggested is 400°F/hr per inch (204.4°C/hr per 25 mm) thickness of the vessel, but in no case more than 400°F/hr (204.4°C/hr). The manufacturer can always go for a lower heating rate. However, too low a heating rate may be too expensive.

2.3 Soaking Temperature and Time

For carbon steel vessels, the soaking time and temperature are specified in UCS 56. Soaking temperature is related to the material specification, whereas the time is based on the thickness of the shell of the vessel. For P. No. 1 materials, the soaking time specified is 1 hr per inch (1 hr per 25 mm) of vessel thickness with a minimum of 15 min. Similarly the soaking temperature specified is 1100°F (593.3°C). Soaking at temperatures lower than that specified calls for extended soaking time, as specified in UCS 56.1. However, it is recommended to carry out the PWHT at the specified temperature and time frame so that detrimental effects due to spheroidization can be avoided.

2.4 Cooling Rate

After soaking, the cooling starts. As per UCS 56, the recommended maximum cooling rate is 500°F/hr per inch (260°C/hr per 25 mm) thickness of the vessel but in no case greater than 500°F/hr (260°C/hr). Slower cooling rates are always preferred, the only consideration being cost. The cooling rate specified is only up to 800°F (427°C), below which it may be cooled in still air.

3 HEAT TREATMENT CYCLE

The chart indicating the heating rate, soaking temperature/time, and cooling rate is the heat treatment cycle and shall form a part of the procedure previously approved as per the provisions of the code. All the other details elaborated are the ones that are normally followed by the operator of the furnace.

4 PLACEMENT OF THERMOCOUPLES

The number of thermocouples to be used for monitoring the temperature is calculated based on the diameter, length, and position of other welded attachments on the vessel as per the code. Once the number of thermocouples is finalized based on the code and client specification, the location of the same is fixed based on the configuration of the vessel as well as the heating method proposed and the location of the heat source. By looking at the position of the heat source and the configuration of the vessel, one can easily locate the probable cold and hot spots. Even though the very purpose of the thermocouple is to monitor the temperature, it should also indicate the difference in temperature at the hot and cold spots. Because of this reason, thermocouples shall be randomly provided at the hot and cold spots evenly as far as possible. In case the difference in temperature between thermocouples is found to be in excess of that stipulated in the codes [i.e., 250°F (121°C) while heating and 150°F (65.6°C) during soaking], the same shall be brought within the acceptable limits by either bringing down the heating rate or by providing additional heat sources at the cold spots.

5 CALIBRATION OF RECORDERS AND THERMOCOUPLES

The recorders to be used shall have valid calibration certificates on the date of performance of the PWHT. Further the thermocouples shall also have valid certification as to the chemical composition of the element as this is the only possible calibration for the same.

EMBLEM

POSTWELD HEAT TREATMENTCONFIRMATION REPORT

Order No.		Sheet	of
Purchaser		Attachment to Certificate No	
Description		Date	
Drawing No(s)			

Description of items Post Weld Heat Treated	
Material Specification	

DETAILS OF POSTWELD HEAT TREATMENT

PWHT Chart Number PWHT Method Type of Heating **Heating Rate** Required Actual **Soaking Temperature &Time** Required Actual **Cooling Rate** Required Actual Thermocouple Type Recorder Identification Recorder Chart Speed Colour of Recorder Points Recorder Calibration Details **Hardness Test Results** Type Required Actual Remarks	

Figure 7.1 Postweld heat treatment confirmation report.

6 VERIFICATION OF POSTWELD HEAT TREATMENT AND RECORDS

A certificate indicating the loading temperature, heating rate, soaking temperature, soaking time, cooling rate, unloading temperature, etc., is prepared as in Figure 7.1 for every heat treatment. In addition to this information, the actual heat treatment chart will have the speed of the chart and will accompany this certificate. A sample procedure for PWHT is enclosed in Chapter 15 in which a PWHT cycle is proposed. The actual time/temperature chart obtained from the multipoint recorder connected to thermocouples will reveal the deviation from the proposed cycle. The following aspects shall be verified from the chart:

1. The loading temperature shall be below 800°F (427°C).
2. The heating and cooling rates shall be either equal to or lesser than that specified in the procedure drawn out as per code. For calculating this the chart speed has also to be noted down.
3. Soaking temperature and time shall be marginally above that specified.
4. The unloading temperature shall be less than 900°F (482°C), that specified in the code.
5. All recorders and thermocouples shall have a valid calibration certificate.

8

Hydrostatic Testing

1 GENERAL

Hydrostatic testing is the last test performed in the manufacture of a pressure vessel prior to its delivery to the client. This is to prove the integrity of the vessel under stringent conditions that simulate the design conditions. This being the last test, rework of any type is not permitted on the vessel afterward. Upon completion, the vessel is released only for subsequent work like cleaning, painting, etc., as desired in the purchase order.

2 BILL OF MATERIAL CHECK

The hydrostatic test being the last operation in the manufacturing process, all direct welding to the vessel shall be completed prior to this. Therefore, it is essential to verify that all components shown in the fabrication drawing have been fixed on the equipment. The bill of materials (BOM) is the authentic parts list for the vessel, each and every item of the BOM has to be verified for its physical presence on the equipment. If there are any loose parts or parts that can be welded later on a pad plate (i.e., not directly welded to the shell), they must be listed. For completeness, all the drawings pertaining to the vessel shall be subjected to same verification. The details of this inspection have to be incorporated in the pre–hydrostatic testing (pre-hydro) inspection report enclosed as Figure 8.1.

3 WELDING COMPLETION CHECK

All welds on the vessel have to be inspected visually. The butt welds shall be checked for uniformity in profile, smooth merging with parent metal, reinforcement, undercuts, uneven ripples, etc. The guidelines given in Section 7.15 will be useful in carrying out the inspection. In case these aspects are observed to be beyond the limits specified, then rework by way of welding and dressing/grinding shall be carried out to bring all the parameters within acceptable limits. Similarly all fillet, socket, and corner welds also shall be subjected to the same for its size, contour, evenness, etc. The results of the visual examination of welds shall also be incorporated in the prehydro inspection report.

4 NONDESTRUCTIVE TESTING COMPLETION CHECK

Prior to the hydrostatic test, all NDT has to be carried out satisfactorily. This has to be verified with control charts pro-

PRE-HYDRO INSPECTION CLEARANCE

Report No.		Job No.		Date	

Equipment No.		Description	

Drawing Nos.

1)		2)		3)	
4)		5)		6)	

A BOM Check

A.1 Drawings Checked	1		2		3		4		5		6	

A.2 Remarks:

1	
2	
3	
4	
5	
6	

B Visual Examination

B.1 Butt Welds

Reinforcement	Satisfactory	Re-work
Contour	Satisfactory	Re-work
Under cuts	Satisfactory	Re-work

B.2 Corner/Fillet/Socket Welds

Size	Satisfactory	Re-work
Contour	Satisfactory	Re-work
Under cuts	Satisfactory	Re-work

B.3 Surface

Free from Arc strikes/Tack Metals/Handling Damages/Other Surface Defects	Satisfactory	Re-work

In case of re-works in any of the above three, please specify the nature of work briefly.------------

----

MPT/LPT: Required/Not required Results: Satisfactory/Re-work

C NDT Completion

C.1 Ultrasonic Testing: Complete/Incomplete Results: Satisfactory/Re-work

C.2 Radiographic Testing: Complete/Incomplete Results: Satisfactory/Re-work

C.3 Magnetic Particle Testing: Complete/Incomplete Results: Satisfactory/Re-work

C.4 Liquid Penetrant Testing: Complete/Incomplete Results: Satisfactory/Re-work

D Reinforcement Pad Air Test

Total number of pads to be air tested		Total number of pads air tested	

Authorized Representative

Figure 8.1 Pre-hydro inspection clearance.

vided along with the concerned NDT procedures. An indication to this effect shall also be given in the pre-hydro inspection release upon satisfactory completion.

5 NONCONFORMITY CLEARANCE CHECK

Nonconformance reports (NCRs) raised during the course of fabrication have to be cleared in all respects by the designated personnel and under no circumstances should there be any violation of the code while resolving a NCR. The disposal of NCRs also shall have to be whetted necessarily by the third-party inspection agencies. A format for the NCR is enclosed as Figure 8.2.

6 SURFACE EXAMINATION

Apart from the stated inspections, the entire inside and outside surface of the vessel as well as nozzles, flanges, etc., are to be subjected to visual inspection. In case of arc strikes, tack welds, etc., they have to be dressed; either an MPT or LPT check has to be carried out; and records to this effect have to be maintained. This again to be incorporated in the pre-hydro inspection report.

7 PAD AIR TEST

All reinforcement pads of nozzles shall be air tested at a pressure of about 1.5 kg/cm^2 (22 psi) with soap solution applied all around the weld both inside and outside, including the weld between pad and the shell. This test has to be carried out before the postweld heat treatment (PWHT) as no repair by welding is permitted after PWHT. This test is extremely important in cases of thin walled vessels.

EMBLEM

NONCONFORMANCE REPORT

Order No.		Sheet	of
Purchaser		NCR No.	
Description		Date	
Drawing No(s)			

Drawing No(s) Applicable to NCR	

Description of Nonconformity	

Forwarded to	M (Design)	M (Plg)	M (Prod)	M (QAC)

Recommended Disposition	

Recommended By	M (Design)	M (Plg)	M (Prod)	M (QAC)

Disposition Accepted By	M (Design)	M (Plg)	M (Prod)	M (QAC)

Forwarded for Rework to	M (Plg)	M (Prod)	M (QAC)

Rework Carried out By	M (Plg)	M (Prod)	M (QAC)

Final Acceptance of Rework	M (QAC)

Figure 8.2 Nonconformance report.

8 SUMMARY SHEETS VERIFICATION

Summary sheets for topics like NDT shall also be prepared as given in the sample MDR so that comprehensive checking of all the stipulated jobs can be done meticulously without fail. A sample summary sheet for radiography is enclosed as Figure 8.3.

9 POSTWELD HEAT TREATMENT

The satisfactory completion of PWHT is another important aspect to be verified. When testing of coupon plates is necessary, it is better to carry this out prior to the hydrostatic testing.

10 HYDROSTATIC TEST PROCEDURE

The pressure at which the hydraulic testing is to be carried out will be shown in the drawing. Usually it is 1.3 times the design pressure with temperature correction as applicable. When vessels are tested in vertical position, the pressure at the topmost point shall be at the required value. The following methodology may be followed for conducting of the hydrostatic test. The pressure shall be built up gradually to half the test pressure and allowed to stabilize at that level for about 5 min. Thereafter, the pressure may be increased in steps of $1/10$ of the test pressure, each time allowing a stabilizing time not less than 2 to 3 min. The vessel needs to be kept at the test pressure until the examination of all weld joints is over. However, it is a good general practice to hold the pressure for half an hour as this time is sufficient to carry out the inspection.

11 PREPARATION FOR THE TEST

Horizontal vessels that are mounted on saddles can be tested in that position on the saddles itself, provided the vessel is

EMBLEM

NDT SUMMARY REPORT

Order No.		Sheet	of
Purchaser		Attachment to Certificate No	
Description		Date	
Drawing No(s)			

Sl.No.	Seam Identification	Requirement of NDT	NDT Report Reference			
			RT	UT	MPT	LPT
1	DS1	100%	R. No. 1,4	--	--	--
2	LS1	100%	2	--	--	--
3	LS2	100%	3	--	--	--
4	LS3	100%	5	--	--	--
5	CS1	100%	21,28	--	--	--
6	CS2	100%	22	--	--	--
7						
8						
9						
10						
11						
12						
13						
14						
15						
16						
17						
18						
19						
20						

QA/QC Manager

Figure 8.3 Nondestructive testing summary report.

temporarily anchored against any toppling forces that may arise during the course of the test.

In the case of very large vertical columns, it may not be possible to do the test in that position. For such vessels, the hydrostatic testing is done in a horizontal position on tempo-

EMBLEM

HYDROSTATIC TEST REPORT

Order No.		Sheet	of
Purchaser		Attachment to Certificate No	
Description		Date	
Drawing No(s)			

Test Pressure

Test Temperature

Test Medium

Duration

Pressure Gauge Range

Pressure Gauge/ Recorder Identification

Pressure Gauge Calibration Details

Result

Date of Test

Attachments

Remarks

Figure 8.4 Hydrostatic test report.

rary saddles made for this purpose. The requirement of the number of saddles needed during the test is decided by the design based on the total hydrostatic test weight and its loading.

In all these cases, the vessel shall be completely filled with water such that the building up of the pressure will be very fast. In order to achieve this air vents shall be provided at all high points to prevent air locking.

The pressure gauges may be provided at the topmost point, and a minimum of two shall be provided. Pressure gauges may be connected only after a valve of sufficient rating so that a defective gauge can be removed without obstructing the test. A similar arrangement shall be provided for the pressurizing device also.

12 GAUGES

An indicating type of gauge shall be provided directly on the vessel. As errors are a common feature on pressure gauges, it is always recommended to have two gauges fitted to the vessel. For large vessels, it is recommended to have a pressure recorder. The normal range required for a pressure gauge is double the test pressure, but in no case less than $1\frac{1}{2}$ times or more than 4 times the test pressure. The gauges and recorder shall be calibrated against a deadweight tester or calibrated master gauge and shall have valid calibration on the date of test.

13 RECORDS

A hydrostatic test certificate signed by the witnessing agency along with the pressure time chart (if possible) will form the documentation for the test. A sample format of this certificate is shown in Figure 8.4.

9

Post–Hydrostatic Test Cleaning

1 GENERAL

After hydrostatic testing, the vessel has to be dried and cleaned to a satisfactory level. Any internals removed for facilitating the hydro test have to be fitted back or be sent as loose parts with due information to the concerned.

2 INSIDE INSPECTION

The inside has to be inspected for debris and sediments collected during the hydrostatic testing. The vessel has to be cleaned well using conventional methods or by the use of dry air. All sediments, rust, slag, etc., has to be removed using damp cotton waste.

In the event that any coating has to be applied, this shall be done after the specified cleaning is carried out. As this does not fall under the scope of fabrication, it is not discussed in this write-up.

3 BOXING UP

Upon completion of cleaning, the fittings removed during testing are fitted back. If all internals are in position, the vessel can be cleared for boxing up.

The nozzles where piping is expected are usually left open, whereas manholes will have blind flanges. There is a chance of rainwater entering through these open nozzles, resulting in the corrosion of the interior during storage of the vessel at site while awaiting installation. Therefore, as a good engineering practice all nozzles shall be blanked using temporary blind flanges of at least 5-mm-thick plates with at least four bolts with rubber gasket between them.

In addition a few bags of silica gel shall be hung inside the vessel from some of the nozzles in order to prevent the humidity from acting against the walls of the vessel.

4 PAINTING OF EXTERIOR

In general, manufacturers do only a primer coating just for transportation purpose. Proper cleaning shall be carried out prior to painting. Since the coating itself is a major topic and does not form a part of the manufacturing process, it is not discussed here. Therefore, upon completion of final primer coating of the vessel, the vessel is considered to be ready for dispatch.

5 INSPECTION RELEASE

In order to issue an inspection release certificate (IRC), the complete scope of delivery shall be known. Of those items specified in the order, a majority of the items compose the main vessel itself. However, there may be items to be sent separately. Therefore, a complete list of items which form the ves-

EMBLEM

INSPECTION RELEASE CERTIFICATE

Purchaser		Certificate No.	
Order No.		Drawing No(s)	
Description		Quantity	

Materials Used	
Basic Dimensions	
Details of Inspection	
Tests Carried Out	
Remarks	
Stamping Details	
Date	
Authorized Representative	

Figure 9.1 Inspection release certificate.

sels and the balance loose items (sent separately) is to be prepared and certified by the planning department and countersigned by the production and QA/QC departments. The loose items shall be tagged with part numbers and sent with a separate packing list.

An IRC is prepared based on these packing lists with suitable remarks on status of the job, percentage completion, etc., for the client (see Figure 9.1).

10

Documentation

1 GENERAL

The final documents that are sent along with the equipment to the client are called the manufacturer's data report (MDR). In order to prepare this, the manufacturer shall have a master document pertaining to the vessel that contains all the relevant details of manufacture. This is the working file of the QAC department and shall have the following documents in it:

1. Material test certificate for each pressure part listed in the fabrication drawings.
2. Material identification reports by the in-house QAC as well as appropriate third parties

3. Stagewise inspection reports
4. Welding inspection reports
5. Nondestructive testing (NDT) reports and a summary
6. Postweld heat treatment (PWHT) reports
7. Prehydrostatic test inspection release
8. Subsequent NDT reports (if applicable)
9. Hydrostatic test report
10. Posthydrostatic test cleaning/boxing up clearance
11. Final inspection release certificate (IRC)

Apart from these, all QAC procedures, including the QAP, welding procedure specifications (WPS), procedure qualification records (PQRs), WQT, nondestructive testing (radiographic, ultrasonic, magnetic particle, and liquid penetrant), postweld heat treatment (PWHT), and final hydrostatic testing (HT) procedures shall be compiled separately with due approval of the concerned client/third-party inspection agency. Out of these, only the WPS needs to be incorporated in the MDR.

2 MANUFACTURER'S DATA REPORT DOCUMENTS

The following documents shall constitute the final MDR.

1. The facing sheet (Figure 10.1).
2. Index for the manufacturer's data report (Figure 10.2).
3. Inspection certificate from the third-party inspection agency.
4. Certificate of manufacture or general certificate, as applicable (Figure 10.3 or Figure 10.4).
5. ASME Forms U1, U1A, U2, U2A, U3, and U4, as applicable.
6. Guarantee certificate (Figure 10.5).

Emblem	

QUALITY ASSURANCE AND CONTROL DEPARTMENT

MANUFACTURER'S DATA REPORT

QUALITY ASSURANCE
AND
CONTROL
DEPARTMENT

JOB ORDER NUMBER	
NAME & NUMBER OF EQUIPMENT	
CUSTOMER	
PURCHASE ORDER NUMBER. & DATE	

Figure 10.1 MDR cover sheet.

Emblem	

QUALITY ASSURANCE AND CONTROL DEPARTMENT

MANUFACTURER'S DATA REPORT (INDEX)

JOB ORDER NUMBER	
NAME & NUMBER OF EQUIPMENT	
CUSTOMER	
PURCHASE ORDER NUMBER. & DATE	

INDEX

SL.No.	Description	Page
1	Third Party Inspection Certificate	
2	Certificate of Manufacture or general certificate	
3	ASME Formats U1, U1A, U2, U2A, U3 and U4 as required	
4	Inspection Release Certificate	
5	Guarantee Certificate	
6	Material Summary Report	
7	Material Test Certificates	
8	Welding Procedure Specification Summary with Weld Map	
9	Welding procedure Specifications	
10	NDT Summary with Weld Map	
11	NDT Reports	
12	Air Leak Test Reports	
13	Dimensional Report	
14	Non Conformity Reports	
15	Pre Hydro Inspection Clearance Report	
16	PWHT Confirmation Report	
17	PWHT Charts	
18	Calibration Reports of Recorders & Thermocouples	
19	Hydrostatic Test Report	
20	Calibration Reports of Pressure Gauges and Recorders	
21	Rub Off / Name Plate Details	

Figure 10.2 MDR index.

Emblem	

QUALITY ASSURANCE AND CONTROL DEPARTMENT CERTIFICATE OF MANUFACTURE

Purchaser		Certificate Number	
		Date	
Order Number		Order Status	
Inspection Date (First)		Inspection Date (Final)	

WELDED PRESSURE VESSEL	Item No. & Description

Design Code or Standard	
Size	
Job Order No.	
Drawing No(s).	
Design Pressure & Temperature	
Type of Construction	
Method of Support	
Internal Equipment	
Non-Destructive Tests	
Compensating Pad(s) Air Test	
Postweld Heat Treatment	
Pressure Test, Medium and Duration	
Design & Drawing Approval	
Remarks	
Stamping Details	
Attachments	
Authorized Representative	

Figure 10.3 Certificate of manufacture.

Emblem	

QUALITY ASSURANCE AND CONTROL DEPARTMENT
GENERAL CERTIFICATE

Purchaser		Certificate Number	
		Date	
Order Number		Order Status	
Inspection Date (First)		Inspection Date (Final)	

Authorized Representative

Figure 10.4 General certificate.

Emblem	

QUALITY ASSURANCE AND CONTROL DEPARTMENT

GUARANTEE CERTIFICATE

Purchaser		Sheet	1 of 1
		Attachment to Certificate No.	
Order Number			
Description		Quantity	
Drawing No(s).			

We hereby guarantee the above product(s) against any manufacturing defects or poor workmanship for a period of ---------------- ( ) months from the date of supply. In the event the product(s) supplied by us fail or develop defects during the above period due to the above reasons, the same shall be rectified by us free of cost.

Date

Authorized Representative

Figure 10.5 Guarantee certificate.

7. Material summary report with an index to materials test certificate (MTC) (Figure 10.6).
8. Welding procedure specification summary with a weld map (Figure 10.7).
9. Welding procedure specifications.
10. NDT summary with weld map.
11. NDT reports.
12. Pad air test reports.
13. Calibration reports of pressure gauges used for pad air test.
14. PWHT reports and charts.
15. Calibration reports of recorders and thermocouples.
16. Hydrostatic test reports.
17. Calibration reports of pressure gauges and recorders.
18. Rub off of stamping on the vessel endorsed by the third-party inspection agency (Figure 10.8).
19. Inspection release certificate (Figure 10.9).

3 MASTER DOCUMENT

When a pressure vessel is manufactured, the QAC department shall have the following original documents for making the MDR:

1. The original material test certificates of all pressure parts and those attached directly to the vessel
2. Original approved QAP, ITP, WPS, PQR, WQT, NDT procedures, PWHT procedures, HT procedures
3. All original inspection details like material identification, fit-up, welding, etc.
4. NDT reports
5. Pad air test reports
6. PWHT reports/calibration reports
7. Hydrostatic test report/calibration reports

EMBLEM

MATERIAL QUALITY REPORT

Order No.		Sheet	of
Purchaser		Attachment to Certificate No	
Description		Quantity	
Drawing No(s)		Date	

Part No.	Description	Thk (mm)	Material Specification	Manufacturer	Identification	Mill/Work Certificate

Figure 10.6 Material quality report.

EMBLEM

WELDING PROCEDURE SPECIFICATION SUMMARY

Order No.		Sheet	of
Purchaser		Attachment to Certificate No	
Description		Quantity	
Drawing No(s)		Date	

Sl.No.	Description	Seam Identification	Applicable WPS	Remarks
1	Seam on Dish 1	DS 01	Xxxx	
2	Shell Long Seams	LS 01 to 06		
3	Long Seams on Cone	LS 07& 08		
4				
5				
6				
7				

Figure 10.7 Summary of welding procedure specifications.

8. Rub off
9. Inspection release certificate.

Apart from these the QAC department has many other documents which will be dealt with separately.

4 ASME CODE FORMS

For ASME U stamp vessels, it is mandatory to fill out Form U1 or U1A, U2, U2A, U3, or U4, as specified in UG-120. These forms shall be used to register the vessel by filing it with National Board of Pressure Vessel Inspectors, 1055 Crupper Avenue, Columbus, Ohio 43229. These formats are given in Appendix W of ASME Section VIII Div (1). The instructions for filling out the forms are given in Table W3 and Table W3.1 of

EMBLEM

RUB OFF / NAME PLATE DETAILS

Order No.		Sheet	of
Purchaser		Attachment to Certificate No	
Description		Quantity	
Drawing No(s)		Date	

RUB OFF OF STAMPING ON MANWAY BODY FLANGE

NAME PLATE COPY

Figure 10.8 Details of rub off/name plate.

EMBLEM

INSPECTION RELEASE CERTIFICATE

Purchaser		Certificate No.	
Order No.		Drawing No(s)	
Description		Quantity	

Materials Used Basic Dimensions Details of Inspection Tests Carried Out Remarks Stamping Details Date Authorized Representative	

Figure 10.9 Inspection release certificate.

Appendix W. A sample form U4 is shown in Appendix W Fig W 3.1.

5 RETENTION OF RECORDS

All the originals pertaining to the above equipment shall be maintained by the QAC department for a period of 5 years, and all RT films shall be kept for a period of 2 years. After this they can be disposed of at the discretion of the manufacturer. However, a brief summary shall be maintained in the databank.

11

Radiographic Testing Procedure for Pressure Vessels

1 SCOPE

The following test procedure is written in accordance with ASME Section V Article 2, "Radiographic Examination," for Pressure Vessels fabricated under ASME Section VIII Div (1).

2 QUALIFICATION OF PERSONNEL

2.1 Nondestructive testing personnel shall be qualified in accordance with the company's written practice established in compliance with SNT-TC-1A of ASNT, "Personnel Qualification and Certification in Non-Destructive Testing."

2.2 Radiographers certified to RT Level I, under the supervision of RT Level II radiographers, may perform radiographic testing.

2.3 RT Level II or III radiographers or their designates may interpret radiographs.

2.4 Radiographers shall be licensed by the Atomic Energy Regulatory Board or other statutory authority of the country to perform RT.

3 EQUIPMENT

3.1 Industrial x-ray units or radioactive isotopes may be used for radiography depending on requirement.

3.2 For normal thickness in the range of 10 to 75 mm of steel, x-ray machines with 350 kV and 6 mA and an iridium-192 isotope source will produce satisfactory results.

3.3 Andrex, Rich Siefert, Gilardoni, and Siemens are some of the good makers of x-ray machines. Tech Ops, Teletron, and Amersham are some of the good makers of gamma ray sources.

3.4 The size of the focal spot for the x-ray machine and the size of the source shall be capable of meeting the geometric unsharpness limitation as specified in T-285.

4 SURFACE PREPARATION

4.1 *Materials*. The surface shall satisfy the requirements of applicable material specification with additional conditioning, if necessary, by any suitable process to a degree that surface irregularities cannot mask or be confused with discontinuities.

4.2 *Welds*. The weld ripples or weld surface irregularities on both the inside (wherever accessible) and outside shall be removed by any suitable process to such a degree that the resulting radiographic image, due to any irregularities, cannot mask or be confused with the image of any discontinuity.

4.3 *Surface finish.* The finished surface of all butt welds may be flush with the base metal or may have reasonably uniform crowns, with reinforcement not to exceed that specified in clause UW35 of ASME Section VIII Div (1).

5 DETECTION AND PREVENTION OF BACK SCATTER

5.1 As a check of the back-scattered radiation, a 12-mm lead letter B of 1.5 mm thickness is attached to the back of the film holder during RT. If a light image of the B appears on the radiograph, it indicates that back-scattered radiation is excessive.

5.2 Wherever practical, a 2- to 3-mm-thick lead sheet or additional layer of intensifying screen will be placed at the back of the film holder during radiographic exposure to shield the film from scattered radiation.

6 SYSTEM IDENTIFICATION

6.1 A system shall be devised to produce permanent identification on the radiograph traceable to the contract, component, and weld or part numbers as appropriate. In addition, the manufacturer's name and the date of the radiograph shall be included. In no case is the identification provided to interfere with the weld under examination.

7 THICKNESS RANGE

7.1 The recommended thickness of steel and steel alloys for the sources is as follows.

Source	Steel thickness
X-ray generator 350 kV & 6mA	10 to 80 mm
Iridium-192 gamma ray source	10 to 75 mm

However, this limitation may be extended when the radiographic technique used demonstrates that the required radiographic sensitivity has been obtained or when agreement has been obtained from all parties concerned.

8 FILM AND SCREEN TYPE

8.1 Radiographs shall be made using industrial radiographic films of type 2 (i.e., fine grain, high contrast). Examples are Kodak AA, Agfa Geavert D7, Fuji 100, and Dupont 65. However, when type 2 films fail to achieve the required sensitivity for thinner materials, type 1 films may be used.

8.2 All unexposed films shall be stored in a clean dry place so that they are protected from the effects of heat, undue pressure, radiation, light, and damaging fumes or vapors.

8.3 Lead foil intensifying screens, each of 0.125 mm thickness shall be placed on either side of the film in close contact to the film and shall be loaded in light-proof film holders.

8.4 Lead screens shall be free from scratches, dirt, dents, and moisture, which may impair the quality/sensitivity of the film or induce false indication/image on processed films.

9 RADIOGRAPHIC TECHNIQUE

9.1 A single-wall exposure technique shall be used for radiography whenever practical. In cases where this is not possible, a double-wall technique may be used. Adequate number of exposures shall be made to demonstrate that the required coverage has been obtained.

9.2 In the single-wall technique, the radiation passes through only one wall of the weld that is viewed for acceptance on the radiograph.

9.3 In the double-wall technique, there are two options based on the size of the object being radiographed.

1. In single-image viewing, the radiation passes through the two walls but the image of the weld on the film-side wall is viewed for acceptance on the radiograph. When complete coverage is required for the circumferential seam weld, a minimum of three exposures shall be taken at 120° apart.
2. For materials and welds in components of 80 NB or less in nominal diameter, a technique may be used in which the radiation passes through the two walls and the weld in both the walls is reviewed for acceptance on the same radiograph. For double-image viewing, only source-side penetrameter alone shall be used. When complete coverage is required, a minimum of two exposures shall be taken at 90° to one another.

10 RADIOGRAPHIC PROCEDURES

The following procedures are applicable for gamma ray radiography, which is most commonly used in pressure vessel manufacture. These procedures are capable of meeting all the requirements specified in the referencing code if followed meticulously.

10.1 Procedure Number RT 01

Component	Girth butt weld in pipes
Material	Carbon steel/alloy steel/stainless steel
Diameter	Above 80 NB up to 150 NB
Thickness	Any pipe schedule
Welding process	Fusion welding
Equipment	Tech-Ops Model 660 with remote control
Radiation source	Irirdium-192
Source strength	5 to 50 Ci
Source size	2 mm × 2 mm
Film type	ASTM Type II or better
Intensifying screen	Lead screen 0.125 mm front and back
Radiographic technique	Double-wall single-image
Source-to-film distance (SFD)	Pipe outside diameter
Number of exposures	Minimum three exposures at 120° apart and four for schedule 160 or xxs pipes
Image quality indicator (IQI)	As per Table T-276
IQI position	Film or source side
Density	2.0 to 4.0
Sensitivity	As per Table T-276

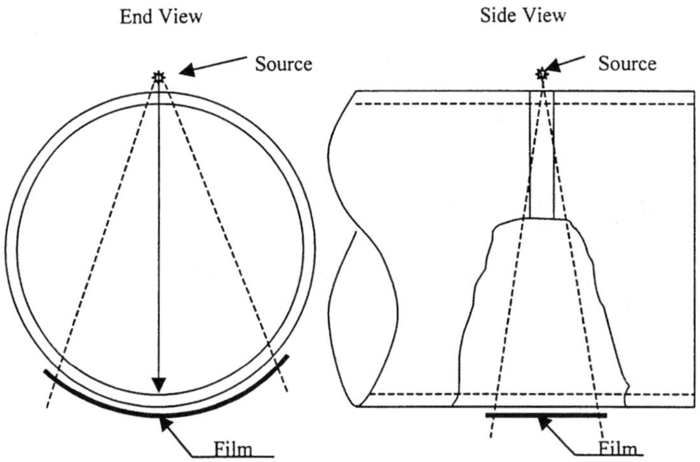

10.2 Procedure Number RT 02

Component	Girth and longitudinal butt weld in pipes
Material	Carbon steel/alloy steel/stainless steel
Diameter	200 NB or greater
Thickness	Any pipe schedule
Welding process	Fusion welding
Equipment	Tech-Ops Model 660 with remote control
Radiation source	Iridium-192
Source strength	5 to 50 Ci
Source size	2 mm $\times$ 2 mm
Film type	ASTM Type II or better
Intensifying screen	Lead screen 0.125 mm front and back
Radiographic technique	Double-wall single-image
SFD	Pipe outside diameter
Number of exposures	Minimum three exposures at 120° apart
IQI	As per Table T-276
IQI position	Film or source side
Density	2.0 to 4.0
Sensitivity	As per Table T-276

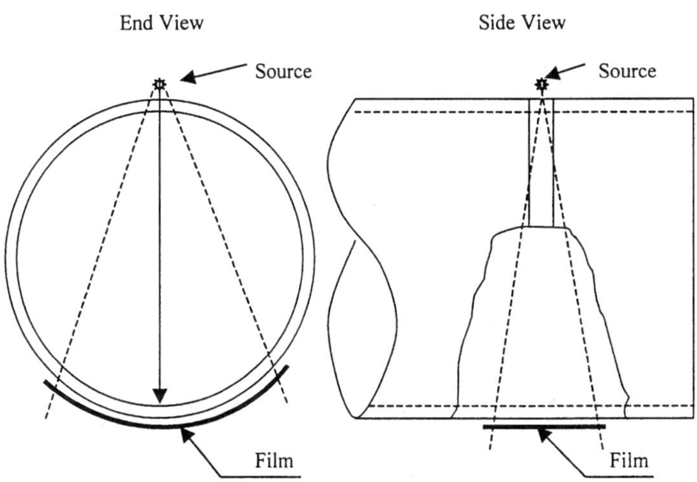

10.3 Procedure Number RT 03

Component	Girth butt weld in pipes
Material	Carbon steel/alloy steel/stainless steel
Diameter	600 NB or greater
Thickness	10 mm and above
Welding process	Fusion welding
Equipment	Tech-Ops Model 660 with remote control
Radiation source	Iridium-192
Source strength	5 to 50 Ci
Source size	2 mm × 2 mm
Film type	ASTM Type II or better
Intensifying screen	Lead screen 0.125 mm front and back
Radiographic technique	Single-wall single-image (panoramic)
SFD	Radius of pipe
Number of exposures	Single exposure
IQI	As per Table T-276
IQI position	Film or source side
Density	2.0 to 4.0
Sensitivity	As per Table T-276

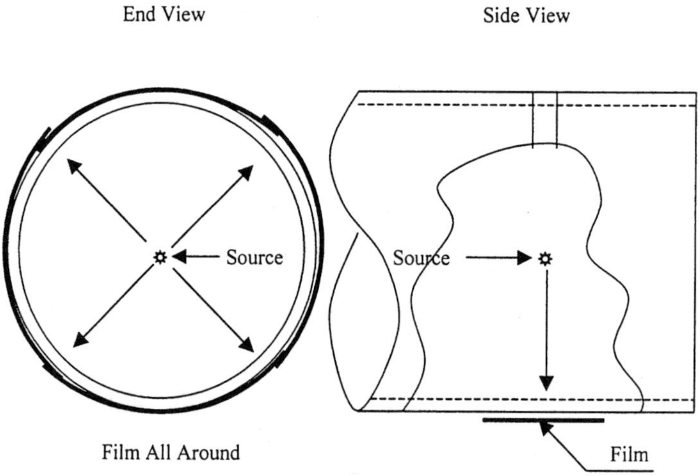

10.4 Procedure Number RT 04

Component	Girth and Longitudinal butt weld in pipes
Material	Carbon steel/alloy steel/stainless steel
Diameter	600 NB or greater
Thickness	10 mm and above
Welding process	Fusion welding
Equipment	Tech-Ops Model 660 with remote control
Radiation source	Iridium-192
Source strength	5 to 50 Ci
Source size	2 mm × 2 mm
Film type	ASTM Type II or better
Intensifying screen	Lead screen 0.125 mm front and back
Radiographic technique	Single-wall single-image
SFD	Minimum of 300 mm
Number of exposures	Single exposure
IQI	As per Table T-276
IQI position	Film or source side
Density	2.0 to 4.0
Sensitivity	As per Table T-276

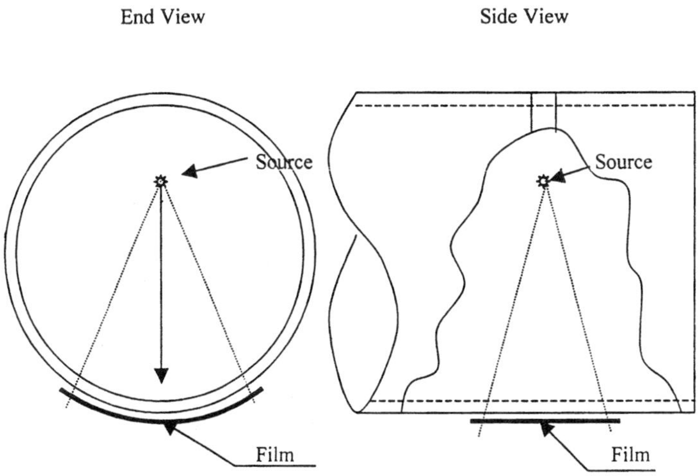

10.5 Procedure Number RT 05

Component	Girth butt weld in small diameter pipes
Material	Carbon steel/alloy steel/stainless steel
Diameter	65 NB or less
Thickness	Schedule 160
Welding process	Fusion welding
Equipment	Tech-Ops Model 660 with remote control
Radiation source	Iridium-192
Source strength	5 to 50 Ci
Source size	2 mm × 2 mm
Film type	ASTM Type II or better
Intensifying screen	Lead screen 0.125 mm front and back
Radiographic technique	Double-wall single-image (elliptical)
SFD	Minimum of 300 mm
Number of exposures	Minimum two exposures at 90° to one another
IQI	As per Table T-276
IQI position	Film or source side
Density	2.0 to 4.0
Sensitivity	As per Table T-276

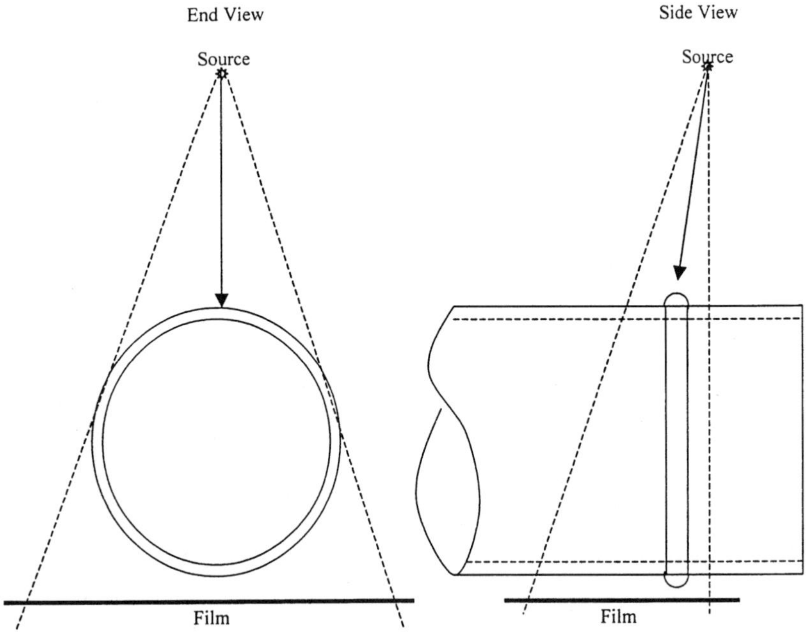

10.6 Procedure Number RT 06

Component	Girth butt weld in small diameter pipes
Material	Carbon steel/alloy steel/stainless steel
Diameter	65 NB or less
Thickness	Schedule 160
Welding process	Fusion welding
Equipment	Tech-Ops Model 660 with remote control
Radiation source	Iridium-192
Source strength	5 to 50 Ci
Source size	2 mm × 2 mm
Film type	ASTM Type II or better
Intensifying screen	Lead screen 0.125 mm front and back
Radiographic technique	Double-wall double-image (superimposed)
SFD	Minimum of 300 mm
Number of exposures	Minimum three exposures at 120° to one another
IQI	As per Table T-276
IQI position	Film or source side
Density	2.0 to 4.0
Sensitivity	As per Table T-276

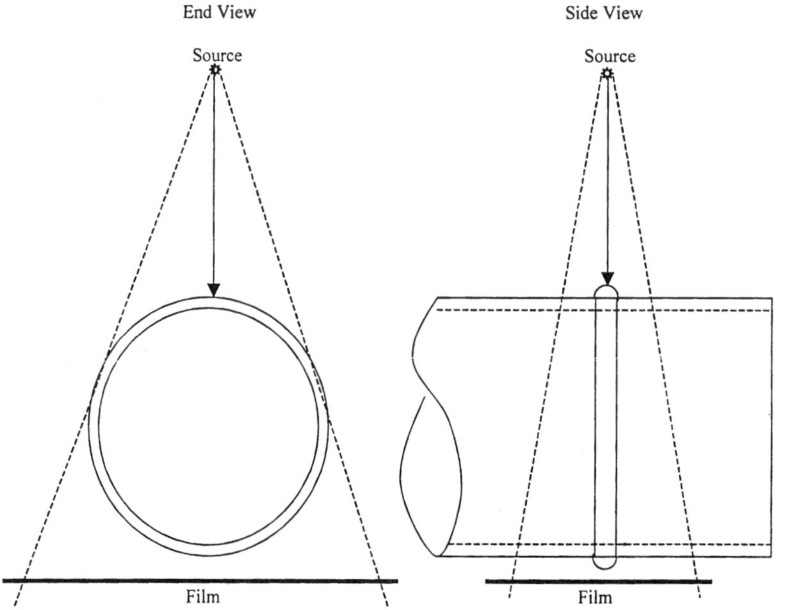

11 GEOMETRIC UNSHARPNESS

The focus–film distance necessary to reduce geometric unsharpness to a negligible amount depends upon the film or film screen combinations, focal spot size, and object film distance:

$$\text{Geometric unsharpness (Ug)} = \frac{Ft}{do}$$

where

F = size of radiation source
t = specimen thickness where the specimen is in contact with film
do = source object distance

When required by the referencing code, geometric unsharpness of the radiograph shall not exceed the values given below.

Material thickness (mm)	Maximum Ug (mm)
Under 50 mm	5.0
50–75 mm	7.5
75–100 mm	10.0
Greater than 100 mm	17.8

12 DIRECTION OF RADIATION

The direction of the central beam of radiation should be centered on the area of interest whenever practical.

13 LOCATION MARKERS

Location markers which are to appear as a radiographic image on the film shall be placed on the part, not on the exposure holder/cassette. The location shall be permanently marked on the surface of the part being radiographed when permitted, or on a weld map, in a manner permitting the area of interest on a radiograph to be accurately traceable to its original location on the part for the required retention period of the radiograph.

14 IMAGE QUALITY INDICATORS

14.1 IQI Selection

Penetrameters shall be either hole type or wire type and shall be manufactured and identified in accordance with the requirements in SE 1025 (for hole type) and SE 747 (for wire type) and its appendices. ASME standard penetrameters shall consists of those in Table T233.1 for hole type and those in Table T233.2 for wire type. The penetrameter shall be made of a material radiographically similar to the material under examination. The thickness on which the penetrameter selection is based is the normal single wall thickness plus the estimated weld reinforcement.

14.2 Placement of IQI

The penetrameter(s) shall be placed on the source side of the component being examined. If this is practically inaccessible, the IQI may be placed on the film side with the lead letter F to indicate that the IQI is placed on the film side. This letter F shall in no way interfere with the weld area that is under scrutiny. Equivalent sensitivity between source-side and film-side penetrameters can be established when required.

The penetrameter(s) may be placed adjacent to or on the weld. The identification numbers and, when used, the letter F shall not be in the area of interest, except when geometric configuration makes it impractical. Wire type penetrameters can be placed over the weld so that the length of wires is perpendicular to the weld.

14.3 Number of IQI

For cylindrical vessels or flat surfaces where one or more films are taken in one exposure, at least one penetrameter image shall appear on each radiograph at the end of the diagnostic film length. In the case of the wire type penetrameter, the smallest wire shall be facing the extreme end of the film.

An exception to this shall be for panoramic exposure (a single exposure of a complete circumference); a minimum of three penetrameters shall be placed equally at 120° apart around the circumference. Where sections of additional longitudinal seams are radiographed simultaneously, additional penetrameters shall be placed on each segment of the longitudinal seam being radiographed.

15 QUALITY OF RADIOGRAPH

All radiographs shall be free from mechanical, chemical, or other blemishes to the extent that they do not mask and are not confused with the image of any discontinuity in the area of interest of the object being radiographed. Such blemishes shall include but are not limited to

1. Fogging
2. Processing defects such as water streaks, water marks, or chemical stains
3. Scratches, finger marks, crimps, dirt, static marks, smudges, or tears
4. False indications due to defective screen

16 DENSITY OF RADIOGRAPH

The transmitted film density through the image of the body of the appropriate hole type penetrameter or adjacent to the designated wire type penetrameter and the area of interest shall be 1.8 minimum for single viewing for radiographs made from x-ray source and 2.0 minimum for radiographs made from γ-ray source. For composite viewing of multiple film exposures, each film of the complete set shall have a minimum density of 1.3. The maximum density shall be 4.0 for either single viewing or composite viewing. A tolerance of 0.05 density is permitted for variations between densitometer readings.

If the density of the radiograph varies by more than −15% to +30% for the density through the body of the hole type penetrameter, within minimum and maximum allowable density range, the additional penetrameter shall be used for each exceptional area and the radiograph retaken.

The film density can be checked with a calibrated densitometer or by a visual comparison against film density strip.

17 SENSITIVITY OF RADIOGRAPH

17.1 The radiograph shall have sufficient sensitivity to display the hole penetrameter image and the specified hole, or the designated wire of a wire type penetrameter, which are essential indications of the image quality of the radiograph. The radiographs also shall display the identification numbers and letters. The requirements for this are given in Table T276.

17.2 Alternatively the radiographic sensitivity can also be calculated in the case of wire type penetrameters based on single-wall thickness as follows.

17.3 For objects of transition thickness, the thickness of the thinner section shall be considered.

17.4 IQI Sensitivity = (thickness of the thinnest wire visible) × 100/thickness of the material.

17.5 The following table depicts the expected radiographic sensitivity for typical gamma ray technique.

Specimen thickness (mm)	IQI sensitivity percentage for wire type penetrameters
6	—
12	2.1
25	1.6
35	1.4
50	1.3
75	1.1

18 FILM PROCESSING

18.1 The process considered here is manual processing.

18.2 No more film should be processed than can be accommodated with a minimum separation of 12.7 mm.

18.3 The temperature and time of development shall conform to the manufacturer's recommendations. However the normal development time is 5 to 8 min at 68°F (20°C).

18.4 The film will be loaded in a hanger and slowly immersed in the developer, continuously agitated vertically and horizontally for a few seconds at 1-min interval for approximately 5 min.

18.5 After the development is complete, the films will be immersed in the stop bath acid or by rinsing with vigorous agitation in clear water to discontinue the activity of the developer that remains in the emulsion for at least 30 sec.

18.6 After the stop bath, the films will be immersed in the fixer (films must not touch one another in the fixer) and agitated vertically for at least 10 sec and again at the end of first minute, to ensure uniform and rapid fixation. The films will be kept in the fixer until fixation is complete (that is at least twice the time needed to clear an unexposed film), but not more than 15 min in relatively fresh fixer. Frequent agitation will shorten the fixation time.

18.7 The use of hypoeliminator or fixer neutralizer between fixation and washing may be advantageous.

18.8 The films will be immersed in fresh water and rinsed for at least 5 min to ensure complete removal of the fixer that may remain in the emulsion. The films will be dipped in wetting agent for approximately 15 sec to facilitate quick and even drying of the films.

18.9 The washing efficiency is a function of wash water, its temperature and flow, and the film being washed. Generally washing is very slow below 60°F (16°C). When washing at temperature above 85°F (30°C), care shall be exercised not to leave films in the water for too long.

18.10 The films are then left in the drier or left at ambient temperature for drying.

19 VIEWING OF RADIOGRAPHS

The radiograph shall be examined by diffused light in a darkened room and the illuminated area shall be masked to the minimum required for viewing the radiographic image, the edges of the radiograph shall always be masked.

Sufficient time shall be allowed for the film interpreter's eye to become adapted to the lighting conditions in the viewing room before radiographs are interpreted.

20 DOCUMENTATION

20.1 Interpretation of radiographs shall be recorded on the standard radiographic report format that shall include the following details.

1. Job identification
2. Name of the company
3. Radiographic technique
4. Density
5. Sensitivity
6. Name of the radiographer
7. Date of radiography

20.2 Radiographs of each component with the report

shall be presented to the client in a folder with all necessary information, including the interpretation and signature of the concerned.

20.3 A copy of the report shall be maintained by the company that performs the radiography in their records for future reference.

20.4 A radiographic logbook consisting of records of each job performed with report reference number shall also be maintained by the NDT company for traceability.

21 ACCEPTANCE CRITERIA

Acceptance or rejection of the weld shall be in accordance with ASME Section VIII Div (1).

22 IDENTIFICATION OF RADIOGRAPHS

22.1 Identification for Single-Wall Single-Image or Double-Wall Single-Image (Superimposed)

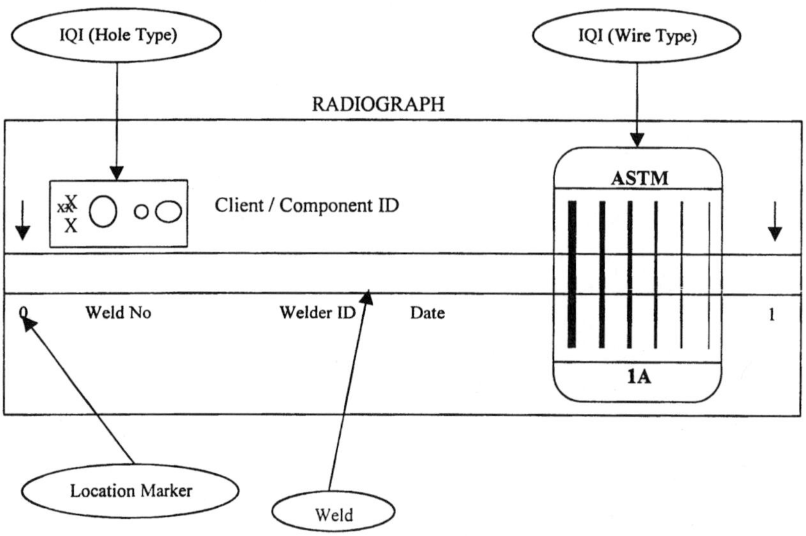

22.2 Identification for Double-Wall Double-Image (Elliptical)

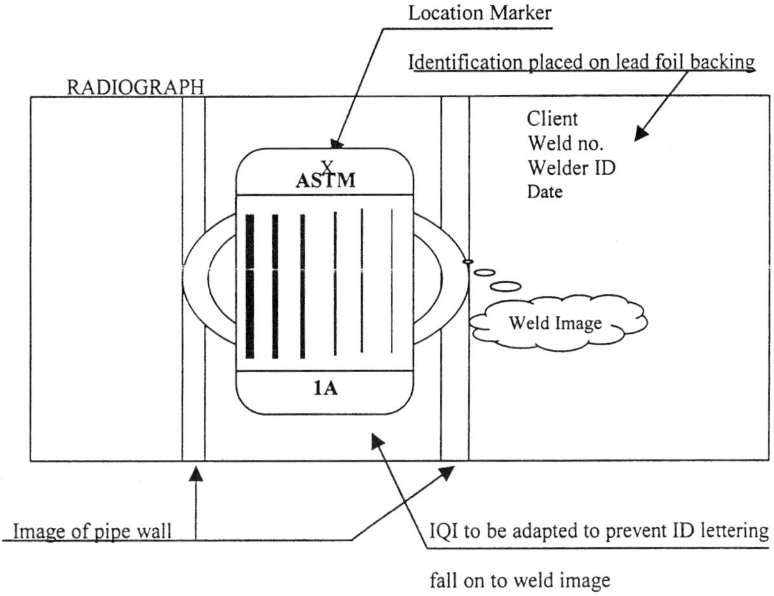

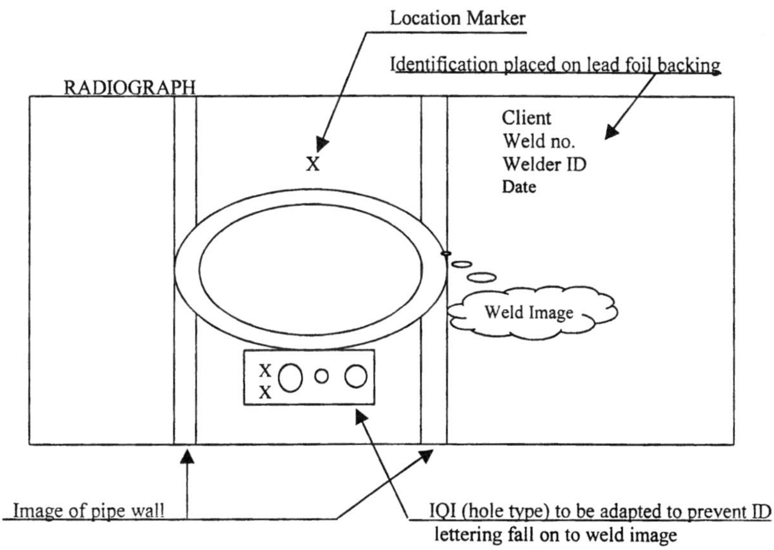

23 REPORT FORMATS

EMBLEM

RADIOGRAPHIC EXAMINATION REPORT

Order No.		Sheet	of
Purchaser		Attachment to Certificate No	
Description		Date	
Drawing No(s)			

Material		Type of Joint	
Welding Process		Surface condition	
Stage of Test		Extent of Test	

X- Ray		Gamma Ray		Material Thickness	
				Film	
kV		Isotope		IQI	
mA		Strength		IQI Placement	
Exposure Time		Exposure Time		Lead Screen: Front	
				: Back	
SFD		SFD		Density	
Focal Spot		Source Size		Sensitivity	

SKETCH

EMBLEM

RADIOGRAPHIC EXAMINATION REPORT- ANNEXURE

Attachment to Certificate No.		Sheet of

Sl.No.	Identification	Date	Film Size	Observation	Remarks

12

Ultrasonic Testing for Pressure Vessels

1 SCOPE

This testing procedure describes the methods and requirements for ultrasonic examination of welds of pressure vessels fabricated under ASME Section VIII Div (1). The procedure is written in accordance with ASME Section V Article 5, Ultrasonic Examination Method for Materials and Fabrication.

This procedure covers ultrasonic examination of plates, forgings, butt welds, set-on, set-in, and set-through nozzle welds.

2 QUALIFICATION OF PERSONNEL

The ultrasonic testing (UT) operator shall be qualified in accordance with the company's written practice established in compliance with ASNT recommended practice SNT-TC-1A, Personnel Qualification and Certification in Non-Destructive Testing.

The UT operators certified to NDT Levels II or III shall perform the test within the scope of their certification, i.e., they shall only perform tests/interpret results on the weld categories (butt welds in pipes and plate, nozzle weld, etc.) for which they are qualified/certified.

3 EQUIPMENT

Necessary equipment shall incorporate the following characteristics:

1. A portable ultrasonic pulse reflection system (A scan presentation), generating pulses of frequency in the range of 1 to 5 MHz will be used for the purpose. Instruments like the Krautkramer USM-2, USK-6, and USK-7 or later versions will provide satisfactory results.
2. Probes of either single crystal or double crystal, round or square, with size ranging from 6.4 to 25.4 mm shall be used. The probe shall produce a second beam (shear wave) in the material being tested within ±2° of the following refracted angles; 45°, 60°, or 70°. For straight probes (0°), a 2.5-MHz frequency is recommended, and that for angled probes is 4.0 MHz.
3. *Screen height linearity*. The ultrasonic instrument shall provide linear vertical presentation within ±5% of the full screen height for 20 to 80% of the calibrated screen height (that is, base line to the maximum calibrated screen points).
4. *Amplitude control linearity*. The ultrasonic instrument shall utilize an amplitude control, accurate over its useful range to ±20% of the nominal amplitude

ratio, to allow measurement of indications beyond the linear range of the vertical display on the screen.

5. *Checking and calibration of equipment*. As a minimum requirement, the proper functioning of the examination equipment shall be checked and calibrated using the calibration standard at the beginning and end of each examination, every 4 hr during examination, or when the examination personnel are changed and at any time when malfunctioning is suspected.

   In any such checks, if any malfunctioning is noticed, all the testing carried out since the last check shall be carried out again using another instrument.

4 REFERENCE AND CALIBRATION STANDARDS

1. *IIW V1 block*. This block is used for general calibration and equipment checks.
2. *IOW block*. This block is used to construct beam profiles and provide a resolution check for shear wave probes.
3. *Distance amplitude correction (DAC) blocks*. These blocks are made per ASME Section V Article 5 clause T 542 for plotting the DAC. The blocks shall be of the same material as that of the test material and meet the requirements of the governing codes and specifications.

5 COUPLANT

Grease, oil, water-based paste (cellulose), or any other suitable liquid may be used as the couplant. The same couplant shall be used in calibration and production testing.

6 SURFACE PREPARATION

The examination surface shall be free from weld spatters, scaling, pitting, irregularities, foreign matter, or thick coating, which could interfere with the ultrasonic wave transmission. When needed, grinding may be done on the surface to be examined to obtain satisfactory and reliable results.

7 GENERAL EXAMINATION REQUIREMENTS

7.1 Examination Coverage

The full volume of the material or the weld shall be examined by moving the search unit (probe) over the examination surface. Each pass of the search unit shall overlap a minimum of 10% of the transducer (piezoelectric crystal) dimension perpendicular to the direction of the scan.

7.2 Rate of Search Unit Movement

The rate of search unit movement for examination shall not exceed 150 mm/sec, unless the calibration is verified at the scanning speed.

7.3 Recording Sensitivity Level

The examination and recording of indications shall be made with respect to the reference level.

8 CALIBRATION

8.1 Basic Calibration Blocks

The basic calibration reflectors shall be used to establish the primary reference response of the equipment. The material from which the block is made shall be of the same or from the same P. number grouping as that of the material under test.

8.2 Basic Calibration Reflectors

The basic calibration reflectors are specified in T 542. For welds on surfaces which are flat or whose diameter is above 500 NB, flat basic calibration blocks as specified in Fig. T 542.2.1 may be used. For welds on curved surfaces whose diameter is less than 500 NB, curved calibration blocks as specified in T 542.3.5 shall be used.

8.3 Distance Amplitude Correction Curve (Primary Reference Level)

1. The search unit shall be positioned for maximum response from the reference block hole, which gives the highest amplitude.
2. The sensitivity control shall be adjusted to provide 80% full screen height indication from the hole. The peak of the indication shall be marked on the screen.
3. The search unit shall be positioned for maximum response from one of the other holes. The peak of the indication shall be marked on the screen.
4. The search unit shall be positioned for maximum amplitude from the third hole and the peak shall be marked on the screen.
5. The search unit shall be positioned for maximum amplitude from the $^3/_4$T hole, after the beam has been reflected from the opposite surface. The peak shall be marked on the screen as the $^5/_4$T position.
6. The screen marks for the side-drilled holes shall then be connected to get the distance amplitude correction curve.

9 TRANSFER LOSS/CORRECTION

The transfer loss is determined by noting down the difference in response received from the reference reflector in the basic calibration block and the production material. The difference is then added to the pregain control setting.

10 SCANNING

10.1 Straight Beam

The scanning of the adjacent base material shall be performed to detect reflectors or laminations that might affect the interpretation using the angle beam examination and is not to be used as acceptance/rejection criteria. Locations and areas of such reflectors are to be recorded in the report.

The scanning shall be performed at a gain setting of at least 2 times the primary reference level. However, the evaluation shall be carried out with respect to the primary reference level only.

10.2 Angle Beam

For longitudinal discontinuities, the angle beam shall be directed at approximately right angles to the weld axis from two directions wherever possible. The search unit shall be manipulated so that the ultrasonic energy passes through the required volume of the weld and adjacent base metal.

The scanning shall be performed at a gain setting of at least 2 times the primary reference level. However, the evaluation shall be carried out at the primary reference level itself.

For transverse discontinuities, the angle beam shall be directed approximately at 15° to the weld axis.

11 DEFECT EVALUATION AND SIZING

Any imperfection that causes an indication in excess of 20% of DAC shall be investigated to the extent that it can be evaluated in terms of the acceptance standards of the referencing code section.

All the reflections from imperfections/defects will be maximized from all angles to find the best angle which is most favorable to the defect orientation.

12 REPORTS AND RECORDS

12.1 Examination Reports

A report of examinations shall be made and recorded in the proper format and shall contain at least the following details:

1. Project title/job number
2. Weld identification
3. UT procedure used
4. Weld geometry
5. Specification for acceptance
6. Equipment used
7. Couplant used
8. Identification and frequency of the probe
9. Surface condition
10. Calibration blocks used
11. Results of the test
12. Name, signature, and qualification level of the operator
13. Date of test

Any defect found shall be recorded with the following details.

1. Location of imperfection
2. Size of imperfection
3. Imperfection type assessment

Reporting of defective areas shall be referred to a datum point which is acceptable to the client.

Copies of approved reports shall be presented to the third-party inspector in order to review and accept reports prior to presentation to the client. Copies of the reports shall be retained by the agency who performs the test.

12.2 Calibration Records

Instrument calibration in accordance with previous sections shall be recorded in proper format.

12.3 Evaluation Record

Record of any evaluation of indications shall be maintained and documented.

13 CLEANING

Cleaning of parts by detergent cleaning or any appropriate cleaning method is required when the couplant will interfere with the subsequent process to be carried out on the part or as required by the client as per applicable specification.

14 ACCEPTANCE CRITERIA

Acceptance or rejection of a component shall be evaluated in terms of the acceptance standard of ASME Section VIII Div (1), namely, Appendix 12.

15 BASIC CALIBRATION BLOCK

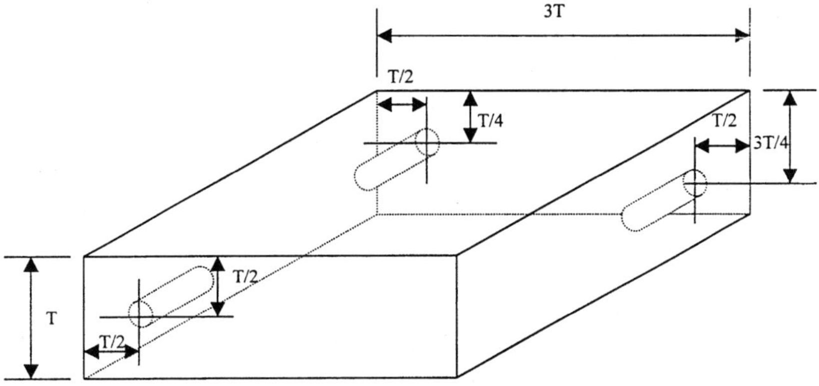

Weld thickness, *t* (in.)	Basic calibration block thickness, *T* (in.)	Hole diameter (in.)
1 or less	3/4 or *t*	3/32
1–2	1 1/2 or *t*	1/6
2–4	3 or *t*	3/16
4–6	5 or *t*	1/4
6–8	7 or *t*	5/16
8–10	9 or *t*	3/8

16 ULTRASONIC TEST PROCEDURE

Test Procedure	UT 01
Component	Butt weld accessible from one side
Weld preparation	Single-V or single bevel
Material thickness	8 mm and above

Sl. no.	Probe	Angle	Size (mm)	Frequency (MHz)
1		0°	10ϕ	2.5 twin crystal
2		45°	8 × 9	4.0 single crystal
3		60°	8 × 9	4.0 single crystal
4		70°	8 × 9	4.0 single crystal

1. Straight probe; scan to one skip distance from the weld toe for lamination.
2. Angle probe; maintain a fixed root scan at a stand-off distance of one-half skip.
3. Angle probe; scan the whole volume of weld with appropriate angle probe.

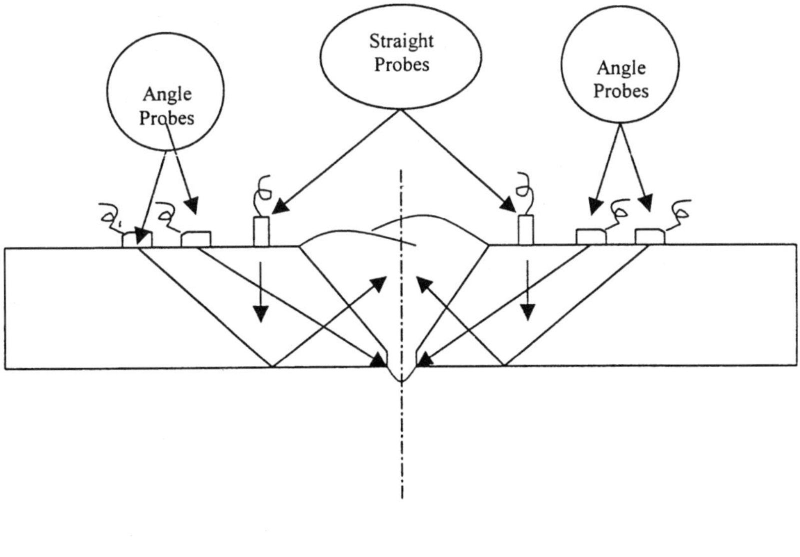

Test Procedure	UT 02
Component	Butt weld accessible from both sides
Weld preparation	Single-V or double-V
Material thickness	8 mm and above

Sl. no.	Probe	Angle	Size (mm)	Frequency (MHz)
1		0°	10ϕ	2.5 twin crystal
2		45°	8 × 9	4.0 single crystal
3		60°	8 × 9	4.0 single crystal
4		70°	8 × 9	4.0 single crystal

1. Straight probe; scan to one skip distance from the weld toe for lamination.
2. Angle probe; maintain a fixed root scan at a stand-off distance of one-half skip.
3. Angle probe; scan the whole volume of weld with appropriate angle probe.

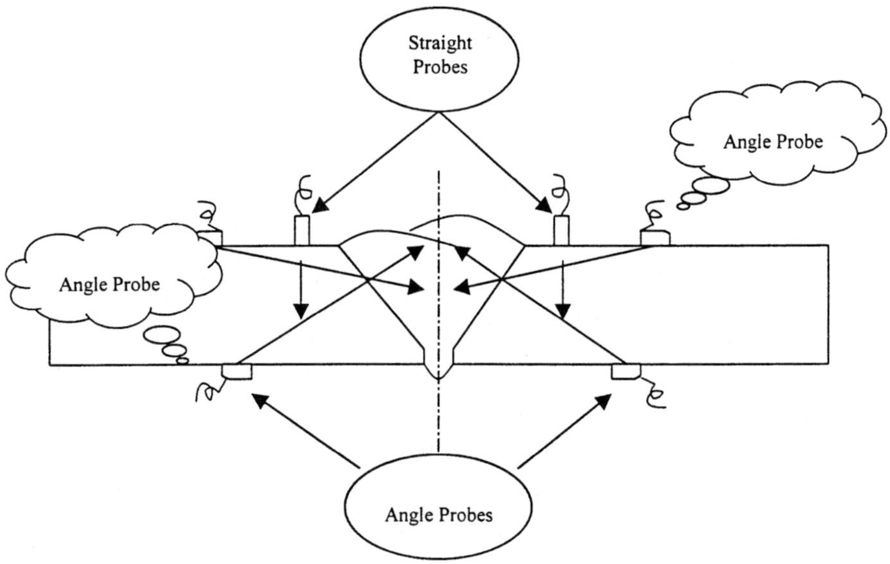

Test Procedure	UT 03
Component	Full penetration set on nozzles
Weld preparation	Single bevel
Material thickness	8 mm and above

Sl. no.	Probe	Angle	Size (mm)	Frequency (MHz)
1		0°	10ϕ	2.5 twin crystal
2		45°	8 × 9	4.0 single crystal
3		60°	8 × 9	4.0 single crystal
4		70°	8 × 9	4.0 single crystal

1. Straight probe; scan to one skip distance from the weld toe on surface A for lamination. Scan on inner shell plate surface B for lack of fusion, cracking, and tearing.
2. Angle probe; maintain a fixed root scan at a stand-off distance of one-half skip.
3. Angle probe; scan the whole volume of weld on surface A with appropriate angle probes.

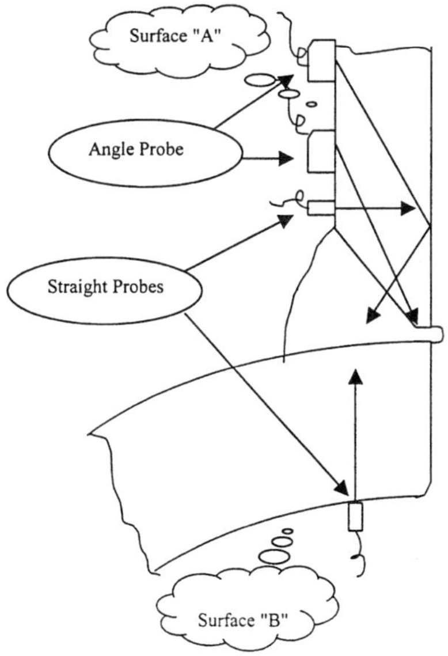

Test Procedure	UT 04
Component	Full penetration set on nozzles
Weld preparation	Single bevel
Material thickness	8 mm and above

Sl. no.	Probe	Angle	Size (mm)	Frequency (MHz)
1		0°	10ϕ	2.5 twin crystal
2		45°	8 × 9	4.0 single crystal
3		60°	8 × 9	4.0 single crystal
4		70°	8 × 9	4.0 single crystal

1. Straight probe; scan to one skip distance from the weld toe on surface A for lamination. Scan on inner shell plate surface B for lack of fusion, cracking, and tearing.
2. Angle probe; maintain a fixed root scan at a stand-off distance of one-half skip.
3. Angle probe; scan the whole volume of weld on surface A and surface C with appropriate angle probes.

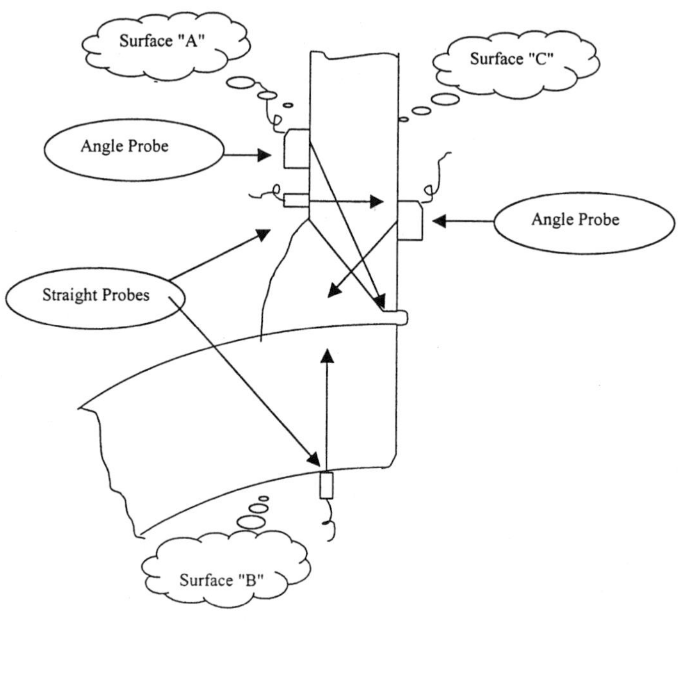

Test Procedure	UT 05
Component	Full penetration set in nozzles
Weld preparation	Single bevel
Material thickness	8 mm and above

Sl. no.	Probe	Angle	Size (mm)	Frequency (MHz)
1		0°	10ϕ	2.5 twin crystal
2		45°	8 × 9	4.0 single crystal
3		60°	8 × 9	4.0 single crystal
4		70°	8 × 9	4.0 single crystal

1. Straight probe: scan to one skip distance from the weld toe on surface A for lamination. Scan on nozzle bore, surface B, for imperfections parallel to the nozzle axis for lack of fusion, cracking in the nozzle wall, and lack of penetration.
2. Angle probe: maintain a fixed root scan at a stand-off distance of one-half skip.
3. Angle probe: scan the whole volume of weld on face A with appropriate angle probes.

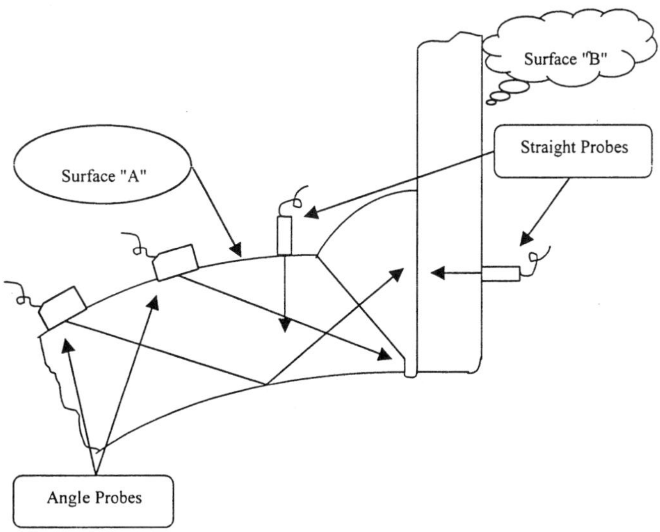

Test Procedure	UT 06
Component	Full penetration set in nozzles
Weld preparation	Single bevel
Material thickness	8 mm and above

Sl. no.	Probe	Angle	Size (mm)	Frequency (MHz)
1		0°	10ϕ	2.5 twin crystal
2		45°	8 × 9	4.0 single crystal
3		60°	8 × 9	4.0 single crystal
4		70°	8 × 9	4.0 single crystal

1. Straight probe: scan to one skip distance from the weld toe on surface A for lamination. Scan on nozzle bore, surface B, for imperfections parallel to the nozzle axis for lack of fusion, cracking in the nozzle wall, and lack of penetration.
2. Angle probe: maintain a fixed root scan at a standoff distance of one-half skip.
3. Angle probe: scan the whole volume of weld on surface A and surface C with appropriate angle probes.

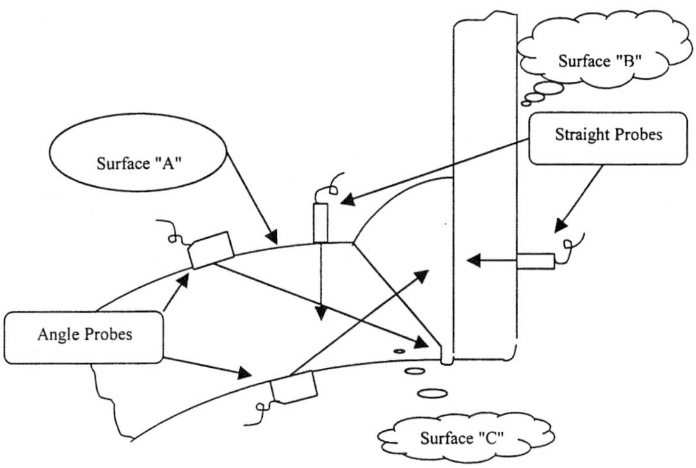

Test Procedure	UT 07
Component	Full penetration set through nozzles (accessible from both sides)
Weld preparation	Double bevel
Material thickness	8 mm and above

Sl. no.	Probe	Angle	Size (mm)	Frequency (MHz)
1		0°	10ϕ	2.5 twin crystal
2		45°	8 × 9	4.0 single crystal
3		60°	8 × 9	4.0 single crystal
4		70°	8 × 9	4.0 single crystal

1. Straight probe: Scan to one skip distance from the weld toe on surface A for lamination. Scan on nozzle bore, surface B, for imperfections parallel to the nozzle axis for lack of fusion, cracking in the nozzle wall, and lack of penetration.
2. Angle probe: Scan the whole volume of weld on surface A and surface C with appropriate angle probes.

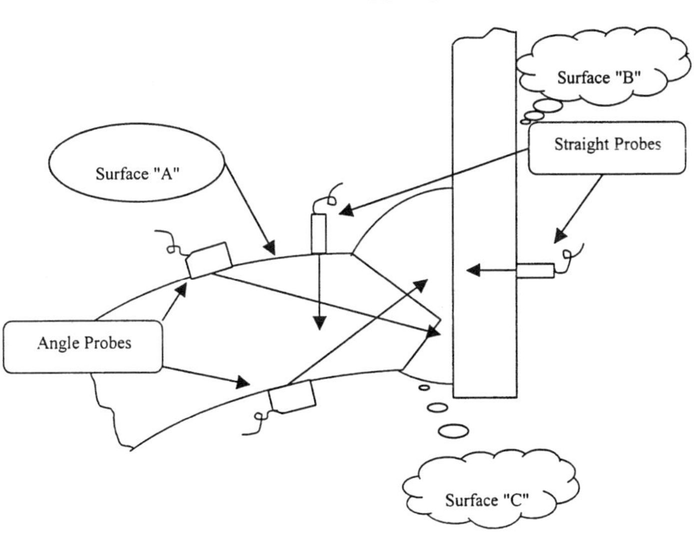

Test Procedure	UT 08
Component	Plate material
Material thickness	100 mm and above

Sl. no.	Probe	Angle	Size (mm)	Frequency (MHz)
1		0°	10ϕ	2.5 twin crystal

1. 0° probe: scan the whole plate from one edge to the other through the grid lines as shown in the sketch.

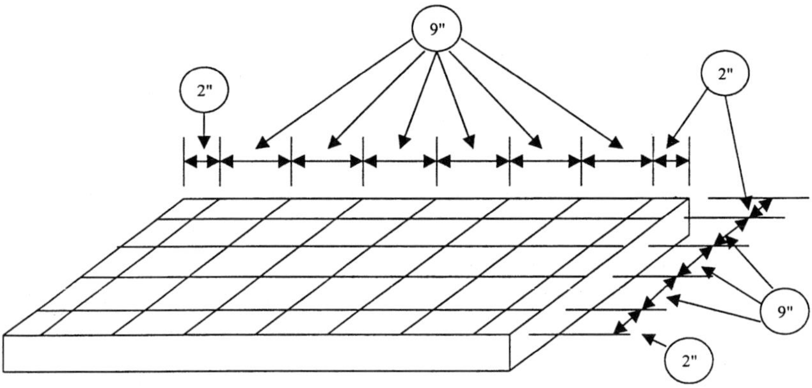

Test Procedure	UT 09
Component	Forging
Material thickness	100 mm and above

Sl. no.	Probe	Angle	Size (mm)	Frequency (MHz)
1		0°	10ϕ	2.5 twin crystal
2		45°	8 × 9	4.0 single crystal
3		60°	8 × 9	4.0 single crystal
4		70°	8 × 9	4.0 single crystal

1. 0° probe and angle probes: scan the whole volume of forging to ensure 100% examination.

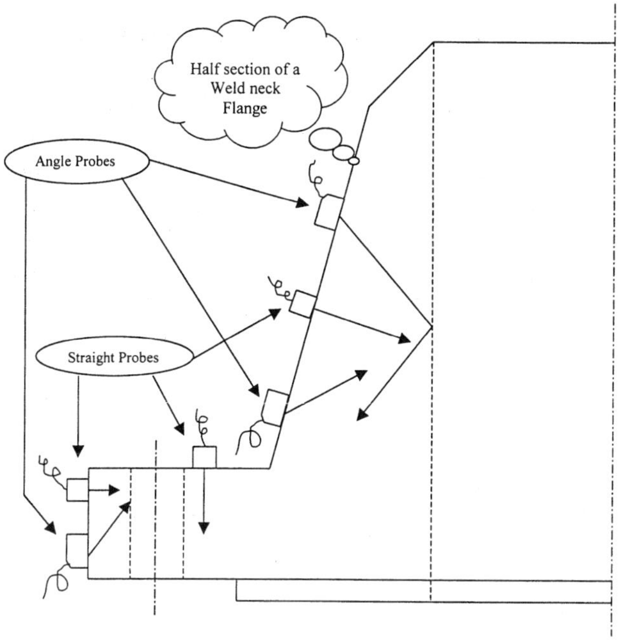

17 DETERMINATION OF INDEX POINT

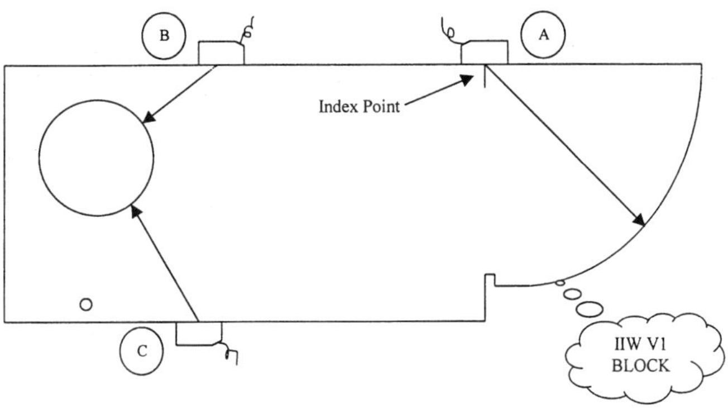

The angle probe is set in position A on IIW V1 block. Maximize the size of the signal from the radius by moving the probe backward and forward parallel to the block edges. The point on the probe, which aligns with the line on the calibration block, is indicative of the point of sound entry (index point).

18 DETERMINATION OF PROBE BEAM ANGLE

The angle probe is set in position B (for angles 45° to 60°) or position C (for angles 60° to 70°) on IIW V1 block, as shown in the figure. The probe is moved back and forth over the line indicative of the probe angle until the signal from the radius of the Perspex is maximized. Compare the index point on the probe with the angle mark on the calibration block V1. The deviation permitted is ±2°.

19 ASSESSMENT TEST—TIME BASE LINEARITY

A compressional wave probe is placed at position D on 25-mm thickness of the IIW V1 block, as shown in the following sketch to obtain a minimum of five back wall echoes in the screen. Adjust the time base so that the first and last back wall echo indication coincides with the appropriate scale marks.

Each indication is adjusted to reference level (80% of full screen height) with the gain for horizontal location examination. The leading edge of each echo should line up, breaking the time base at the appropriate graduation. Record any deviation from the ideal position with relevant echo at approximately half-full screen height. The deviation permitted is 2% of full time base range.

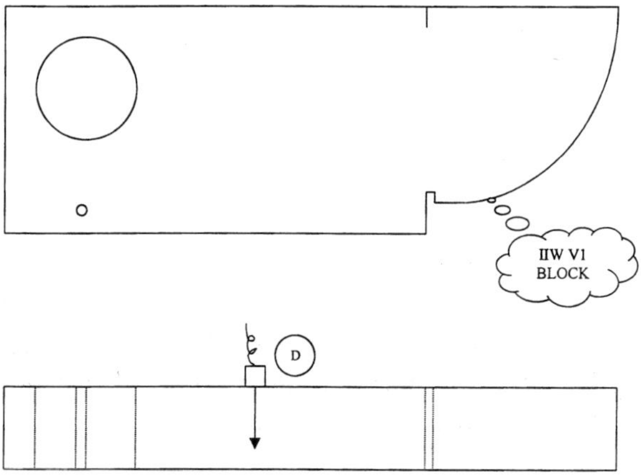

20 ASSESSMENT TEST—SCREEN HEIGHT LINEARITY

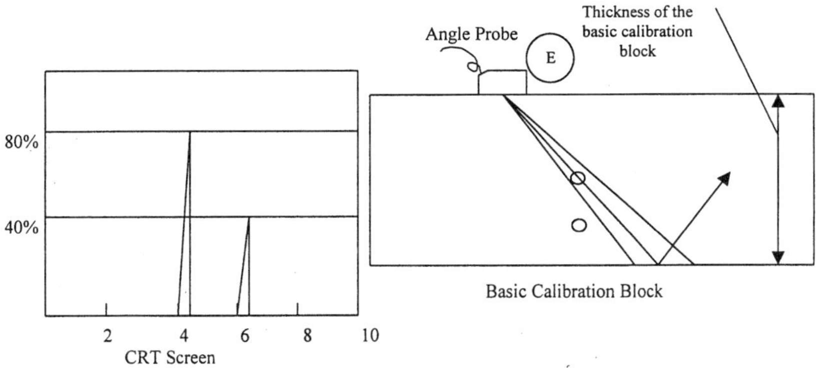

To verify the ability of the ultrasonic instrument to meet the linearity requirement specified earlier, position an angle beam search unit as shown in the sketch so that indications can be observed from both the 1/2 *T* and 3/4 *T* holes in a basic calibration block. Adjust the search unit position to give 2:1 ratio of amplitudes between the two indications, with the larger set at 80% of full screen height. Without moving the search unit, adjust sensitivity (gain) to successively set the larger indication from 100 to 20% of full screen height, in 10% increments (or 2-dB steps if a fine control is not available), and read the smaller indications at each setting. The reading must be 50% of the larger amplitude, within 5% of full screen height. The settings and readings must be estimated to the nearest 1% of full screen. Alternatively, a straight beam search unit may be used on any calibration block which will provide amplitude differences, with sufficient signal separation to prevent over lapping of two signals.

21 ASSESSMENT TEST—AMPLITUDE CONTROL LINEARITY

To verify the accuracy of the amplitude control of the ultrasonic instrument, as required, position an angle beam search unit as shown in the preceding figure so that the indication from the

½ *T* hole in a basic calibration block is at the peak on the screen. With the increase and decrease in attenuation shown in the following table, the indication must fall within the specified limits. Another convenient reflector from any calibration block may be used with angle or straight beam search units.

Indication set at (% of full screen)	dB Control variation	Indication limits (% of full screen)
80%	−6	32 to 48%
80%	−12	16 to 24%
40%	+6	64 to 96%
20%	+12	64 to 96%

The settings and readings must be estimated to the nearest 1% of full screen.

22 CHECK OF RESOLUTION OF FLAW DETECTOR AND PROBE COMBINATION

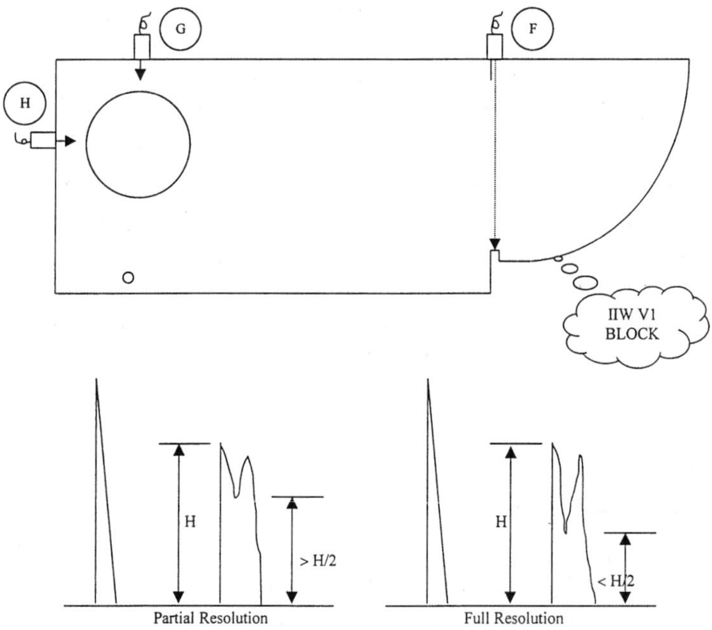

22.1 Compressional Straight Probe

A compressional wave probe is set in position F of IIW V1 block, as shown in the figure, directing the beam toward the depth resolution notch (i.e., toward the three reflectors at distances 85, 91, and 100 mm). The probe is positioned in such a way that the indications at 91- and 100-mm distances from the probe are both at 80% of full screen height.

Measure the echo separation (i.e., echo peak to bottom of the valley between the two echoes) in percentage full screen height. If separation is 50% or more of full screen height, then it is said to be resolved.

Position G and H to determine the near surface resolution or the dead zone of the probe. If the reflection from the Perspex disc at 10 mm (H) and 5 mm (G) could not be displayed, it should be noted as a reflector within the region would not be detected (dead zone).

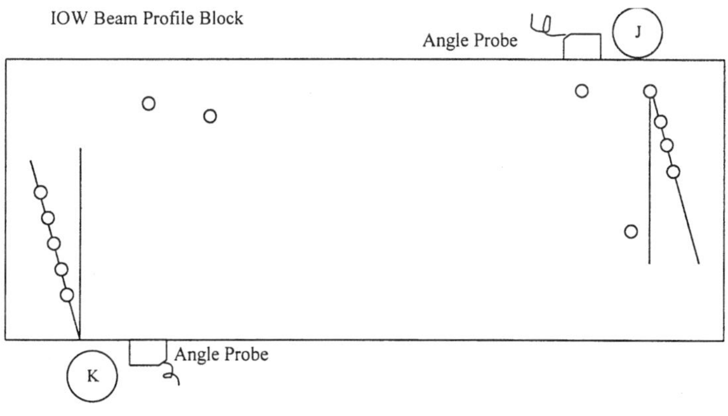

22.2 Shear Wave (Angle) Probe

The Shear wave probe is placed at positions J or K, directing the beam toward a series of drilled holes at different depths. The signal from the drilled hole should be distinguishably displayed.

Near surface resolution can be determined by using a block with drilled holes near the surface.

23 EQUIPMENT QUALIFICATION

Flaw Detector Type: | Date:

Serial Number:

Test	Tolerance			Result	Remarks
Time Base Linearity	±2				
Amplitude Control Linearity	Indication set at % of full screen	dB control change	Indication limits % of full screen		
	80%	-6dB	32 to 48%		
	80%	-12dB	16 to 24%		
	40%	+6dB	64 to 96%		
	20%	+12dB	64 to 96%		
Screen Height Linearity	Difference between smaller signal and 1/2 larger signal height to be better than 5% FSH	Main Signal		Smaller signal	
		100% FSH		FSH	
		80% FSH		FSH	
		60% FSH		FSH	
		40% FSH		FSH	
		20% FSH		FSH	
Equipment resolution	All the three echoes to be clearly resolved	Probes used			
Angle	Marked Angle 45°	Actual Angle		To be within ±2°	
	Marked Angle 60°	Actual Angle			
	Marked Angle 70°	Actual Angle			

General Probe Information

Compression : Type, Crystal Size & Frequency

Shear : Type, Crystal Size & Frequency

Name of Operator : | Signature:

NDT Certification : | Date:

Clients Acceptance :

Remarks

24 PREPARATION OF DISTANCE AMPLITUDE CORRECTION CURVE

The correction curves for angle probes are obtained by using the response from the basic calibration hole. The first point is the PRE. Without changing the gain, the probe unit is then placed at other nodal positions covering the whole examination distance range and the corresponding responses marked on the CRT screen or plotted on graph paper. These points are joined by a smooth line as shown in the following figures. Plotting of the distance amplitude correction (DAC) curve for thickness 25 mm and less is as follows:

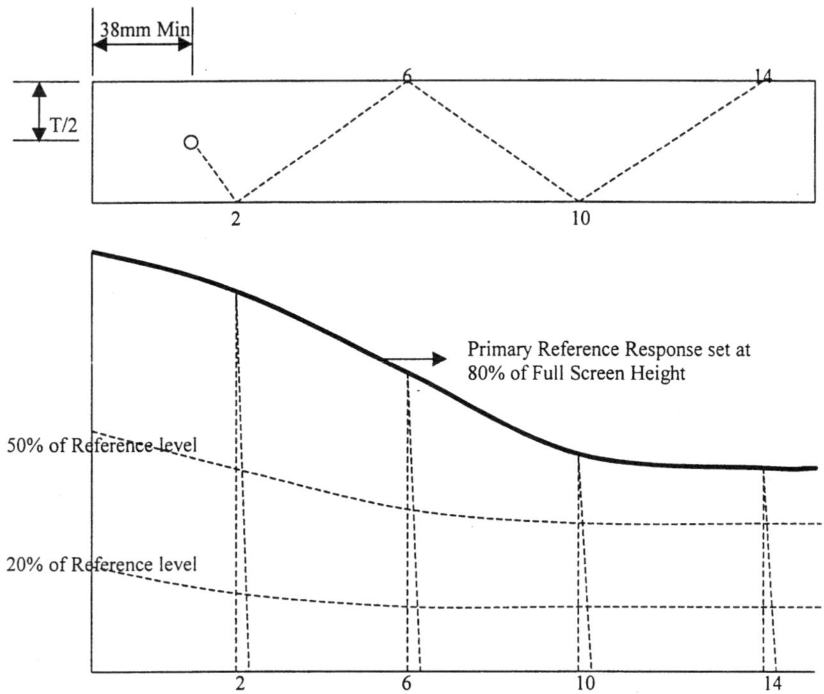

Plotting of the DAC curve for thickness over 25 mm is as follows:

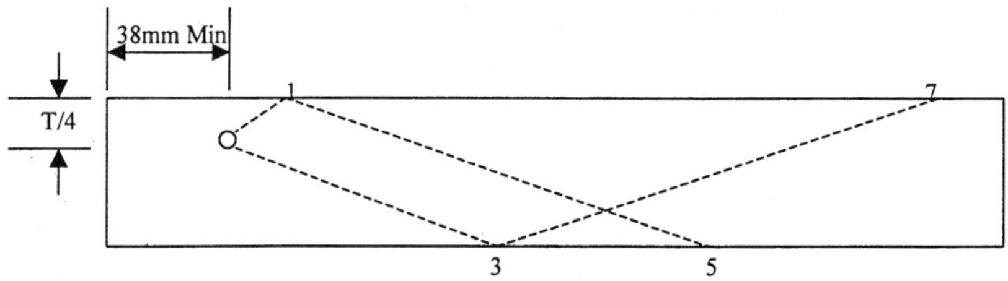

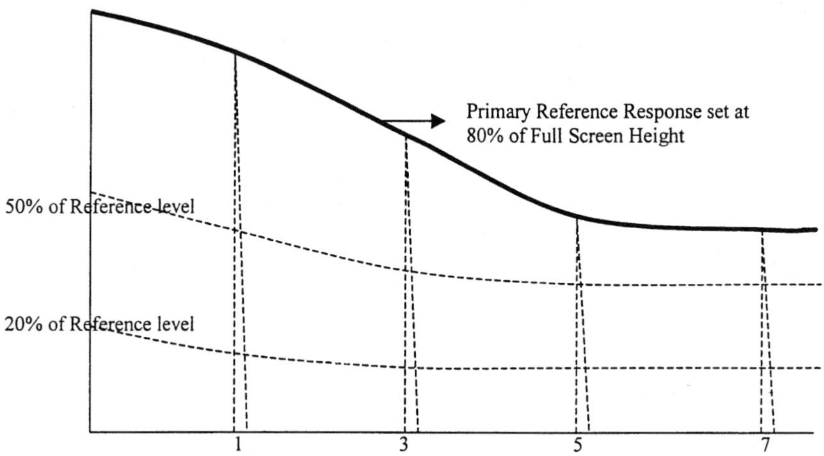

25 TRANSFER CORRECTION

There is a difference in instrument sensitivity between the reference block and test component due to test surface roughness and acoustical attenuation.

1. Use two identical angle probes, one acting as transmitter and the other as receiver.
2. Place the probes in tandem on the broad side of the reference block at one skip distance apart with a weight on top of one of them. They can be kept in a line by the use of a ruler alongside.

3. Peak the signal at 80% full screen height by probe movement and record the signal location on the screen. Note the gain (dB).
4. Reposition the probes to achieve peak signal at two skip distances and join the two points to obtain line R.

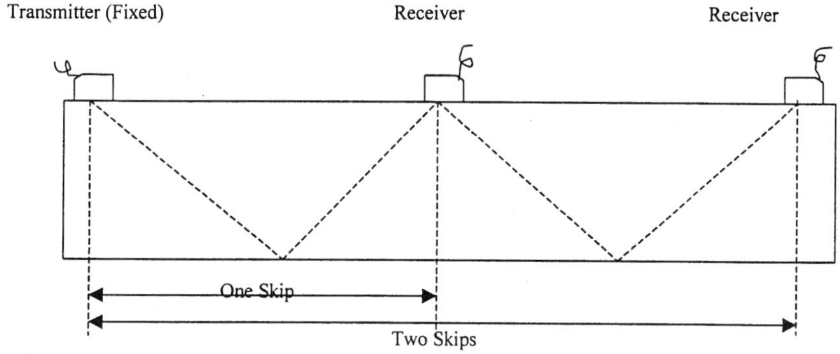

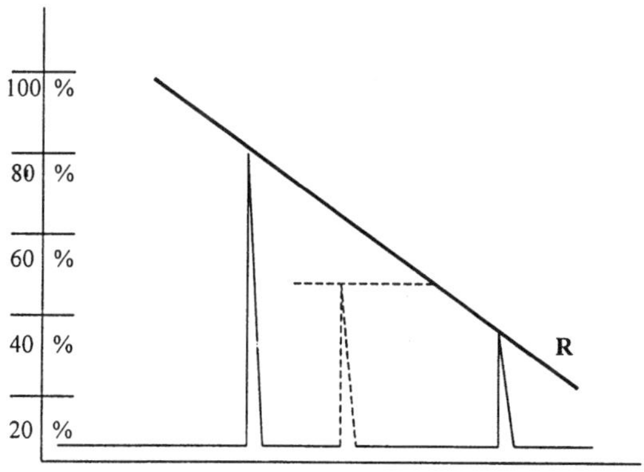

SCREEN DISPLAY

5. Repeat steps 2 and 3 on the test component at the same decibel gain. Increase or decrease the signal where the peak is on line R and note the gain difference.
6. The difference in gain of the instrument in steps 3 and 5 is the transfer correction and has to be added or subtracted during sizing of defects.

26 REPORT FORMATS

EMBLEM

ULTRASONIC EXAMINATION REPORT

Order No.		Sheet	of
Purchaser		Attachment to Certificate No	
Description		Date	
Drawing No(s)			

Material		Probe (1)				
Thickness		Probe (2)				
Surface Condition		Couplant				
Direction of Test		Basic Calibration Block				
Extent of Test		Reference Hole		T		Dia
Test method		Reference Level 80 % FSH			dB	+6dB
Test Equipment		Test range				

SKETCH

EMBLEM

ULTRASONIC EXAMINATION REPORT- ANNEXURE

Attachment to Certificate No.		Sheet of

Scanning surface	Indications		Direction of Measure-ment	Reading Interval (mm)	Stand-Off (mm)	Beam Path (mm)	Type of defect	Remarks
	Sl.No	% Amp						

13

Magnetic Particle Testing for Pressure Vessels

1 SCOPE

This procedure describes the methods and techniques for both dry and wet magnetic particle examination (MPE). This procedure is written in accordance with ASME Section V Article 7, Magnetic Particle Examination of Welds of pressure vessels fabricated as per ASME Section VIII Div (1) requirements.

2 QUALIFICATION OF PERSONNEL

Qualifications required for various personnel are as follows:

1. Personnel performing the magnetic particle inspection shall be qualified in accordance with the com-

pany's written practice established in accordance with ASNT Recommended Practice SNT-TC-1A (Personnel Qualification and Certification in Nondestructive Testing).

2. MT Level II or Level III operators shall carry out magnetic particle inspection and evaluate the acceptability of any magnetic indications.
3. MT Level I operators may carry out magnetic particle testing under the supervision of MT Level II or III personnel and follow approved procedures.
4. MT Level I operators shall not evaluate the acceptability of any magnetic indication.

3 METHOD OF EXAMINATION

Examination shall be carried out by the continuous method; that is, the magnetizing current remains on while the examination medium is being applied and while excess of the examination medium is being removed.

4 MAGNETIZATION TECHNIQUE

The ferromagnetic particles used as an examination medium shall be either wet or dry and may be fluorescent or nonfluorescent.

The combination of an alternating current (AC) electromagnetic yoke (to detect surface discontinuities) and half-wave direct current (HWDC) (to detect subsurface discontinuities) or a permanent magnetic yoke (for detection of surface and subsurface discontinuities) is suitable for localized longitudinal magnetization of small or large components. Permanent magnetic yokes shall only be used when electromagnetic yokes are inapplicable.

5 SURFACE PREPARATION

Satisfactory results are usually obtained when the surfaces are in the as-welded, as-rolled, as-cast, or as-forged condition.

However, surface preparation by grinding or machining may be necessary where surface irregularities could mask indications due to discontinuities.

Prior to magnetic particle examination, the surface to be examined and all the adjacent areas within at least 25 mm shall be dry and free from dust, dirt, grease, lint, scale, welding flux and spatter, oil, or other extraneous matter that could interfere with the examination.

Cleaning may be accomplished using detergents, organic solvents, descaling solutions, paint removers, vapor degreasing, sand or grit blasting, or ultrasonic cleaning methods.

If coatings are left on the part in the area being examined, it must be demonstrated that indications can be detected through the maximum coating thickness applied.

6 EXAMINATION MEDIUM

All magnetic particles used in either the wet or dry MPE technique shall have good contrast color against test surface, high permeability, and low retentivity and shall be sufficiently fine.

Dry powders are heat resistant and are usable at the temperature of the test surface up to 315.6°C (600°F) for large objects with portable magnetizing equipment (yokes and prods) for field testing. Dry powder shall be applied by spraying or dusting directly onto the surface of the part being magnetized.

A powder premix with the suspending vehicle in pressurized cans shall be used for wet magnetic particle method. The temperature of the wet particle suspension and the surface of the part shall not exceed 57.2°C (135°F). The particles carry a very low hazard potential with regard to flammability and toxicity.

With fluorescent particles, the examination is performed using an ultraviolet light, also called a black light. The examination shall be performed as follows.

1. It shall be performed in a darkened area.
2. The examiner shall be in the darkened area for at

least 5 min prior to performing the examination to adapt to dark viewing. If the examiner wears glasses or lenses, they shall not be photosensitive.

3. The black light shall be allowed to warm up for a minimum of 5 min prior to use or measurement taken of the intensity of the ultraviolet light emitted.
4. The black light intensity shall be measured with a black light meter. A minimum of 800 $\mu W/cm^2$ on the surface of the part being examined shall be required. The black light intensity shall be measured at least once every 8 hr. and whenever the work station is changed.

7 EXAMINATION

7.1 Direction of Magnetization

At least two separate examinations shall be performed on each area. During the second examination, the lines of magnetic flux shall be approximately perpendicular to those used during the first examination.

7.2 Examination Coverage

All examination shall be conducted with sufficient overlap to assure 100% coverage at required sensitivity.

8 MAGNETIZING FORCE

8.1 Yoke Magnetizing

The magnetizing force of yokes shall be checked at least once a year or whenever a yoke has been damaged. If a yoke has not been in use for a year or more, a check shall be done prior to its first use.

1. Each alternating current electromagnetic yoke shall have a lifting power of at least 4.5 kg (10 lbs) at maximum pole spacing that will be used.

2. Each direct current or permanent magnetic yoke shall have a lifting power of at least 18.1 kg (40 lbs) at the maximum pole spacing that will be used.

The weight used for verifying the lifting power should be properly stenciled and verified.

9 EVALUATION OF TEST PERFORMANCE AND SENSITIVITY

The overall performance and sensitivity of the magnetic particle system can be tested by

1. Production test parts with known natural discontinuities.
2. Test parts with artificial discontinuities introduced. Example: the test ring specimen with artificial subsurface discontinuities at various depths.
3. *Magnetic field indicators*. When it is necessary to verify the adequacy or direction of the magnetizing field, the magnetic field indicator shall be used by positioning the indicator on the surface to be examined. While using the indicator, a suitable flux or field strength is indicated when a clearly defined line of magnetic particles forms across the copper face of the indicator when the magnetic particles are applied simultaneously with the magnetizing force.

When a clearly defined line of particles is not formed or is not formed in the desired direction, the magnetizing technique shall be changed or adjusted.

10 TEST PROCEDURE FOR YOKE METHOD

1. The use of the magnetic yoke favors the direction of discontinuities where the major axis lies transverse

to the line joining the pole pieces of the yoke. For optimal effectiveness in detecting all types of discontinuities, each area should be tested at least twice, with the lines of flux during one testing approximately perpendicular to the lines of flux during the other.

2. Clean and dry the area of test free from rust, oil, grease, slag, spatter, or any other contaminant that may obstruct the movement of magnetic particles or impair result interpretation. A light grinding to the surface to be examined may be given if found necessary.
3. Apply a thin coat of white contrast paint on the area of interest and allow it to dry prior to the test.
4. Place the yoke on the area of test in a direction where a second application at 90° to the first axis is possible. The distance between the leg span should be maintained between 100 to 200 mm. Ensure good contact of yoke to the test piece. Switch on the current flow (if AC yoke is used).
5. Apply magnetic ink either through aerosol (wet method) or a puffer (dry method) onto the area where the magnetic field has been induced with the current still on.
6. Observe for indications formed by magnetic particles.
7. Cut off the current flow before removing the yoke.
8. Repeat steps 4 to 7 at an axis perpendicular to the first test.
9. When an indication is found, perform step 4 to 7 at an axis perpendicular to the axis of the indication in order to obtain maximum sensitivity.
10. If in doubt, steps 4 to 7 should be repeated and the indication re-evaluated until a result can be obtained.
11. Repeat steps 4 to 10 over the entire area of test.

Maintain an overlap of approximately 50 mm to ensure 100% coverage of the area of test.

12. Remove remaining solvent or iron powder after the test.
13. Record flaw indications found with respect to their location, size, and type.

11 INTERPRETATION OF INDICATIONS

All indications shall be evaluated whether they are relevant, nonrelevant, or false indications. Indications will be revealed by retention of magnetic particles just on top of the defect. All such indications are not necessarily imperfections, however, since excessive surface roughness, magnetic permeability variations (such as at the edge of heat affected zones), etc., may produce similar indications.

An indication is the evidence of a mechanical imperfection. Only indications which have any dimension greater than 1.6 mm ($^1/_{16}$ in.) shall be considered relevant. Two types are as follows:

1. A linear indication is one having a length greater than 3 times the width.
2. A rounded indication is the one of circular or elliptical shape with a length equal to or less than 3 times its width.

Any questionable or doubtful indications shall be re-examined to determine whether or not they are relevant:

1. True (relevant) indications are caused by leakage of fields due to discontinuities in the specimen.
2. Nonrelevant indications are those caused by leakage of fields other than by discontinuities such as from constriction in metal path, change in section, change

in permeability, magnetic writing, or false lines from overmagnetiation.

3. False indications are those that are mechanically held on by gravity, with no relation to leakage fields such as from grease, edge of mill scale, or surface irregularities.

12 POST-TEST CLEANING AND DEMAGNETIZATION

Post-test cleaning and demagnetization of parts magnetized for testing are required when the residual magnetic particles and magnetic field will interfere with the subsequent process to be carried out on the part or as required by the client or specifications.

13 ACCEPTANCE CRITERIA

All indications shall be evaluated in terms of the acceptance standard of ASME Section VIII Div (1) requirement.

14 REPORTING

All tests carried out shall be reported in a format that contains the following information:

1. Name of the client
2. Surface condition
3. Type of equipment used
4. Magnetization technique applied
5. Type of ferromagnetic particles used
6. Identification of components tested
7. Result of interpretation
8. Name and signature or the operator
9. Date of test

15 EXAMINATION OF BUTT WELD USING TRUE CONTINUOUS MAGNETIZATION BY PERMENANT OR AC MAGNETIC YOKE

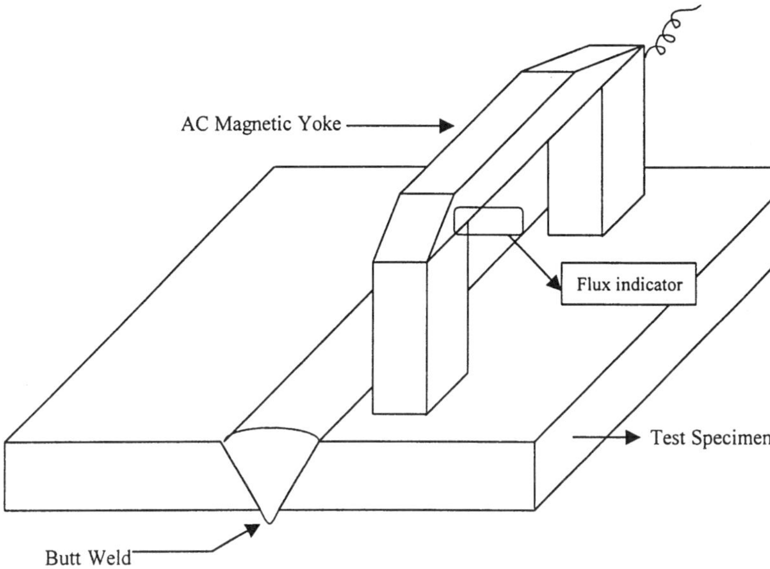

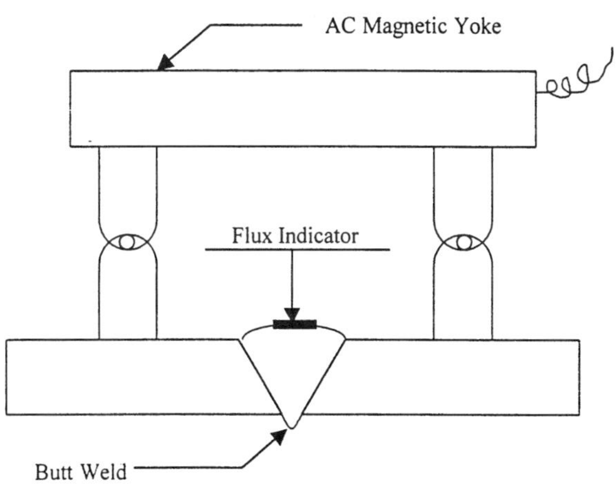

16 EXAMINATION OF FILLET WELD USING TRUE CONTINUOUS MAGNETIZATION PERMENANT OR AC MAGNETIC YOKE

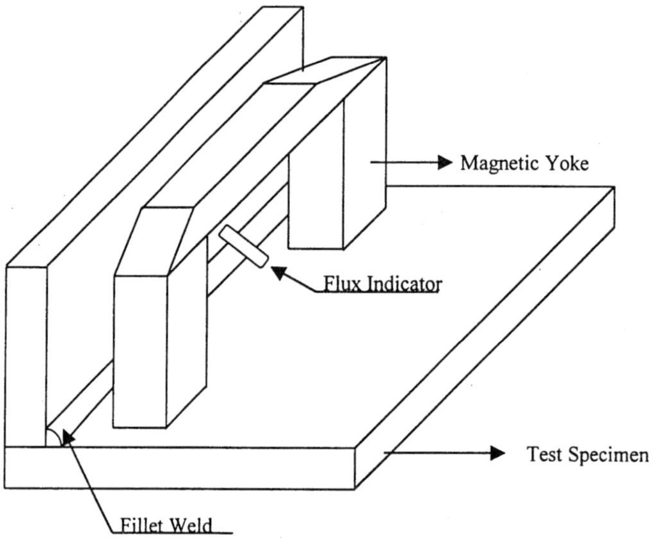

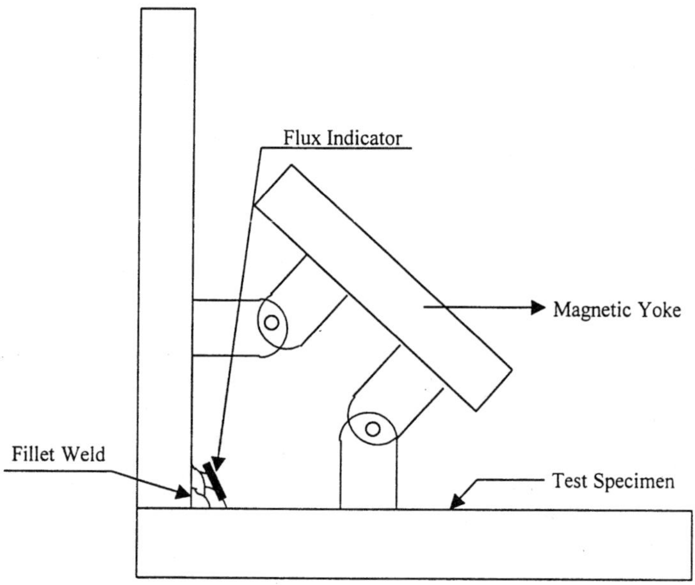

17 REPORT FORMAT

EMBLEM

MAGNETIC PARTICLE EXAMINATION REPORT

Order No.		Sheet	of
Purchaser		Attachment to Certificate No	
Description		Date	
Drawing No(s)			

Equipment		Surface under Examination	
Magnetizing Process			
Magnetizing Current		Surface Preparation	
Technique		Magnetic Particle	
Direction of Field		De-magnetization	
Field Strength		Post-Cleaning	

Sl.No.	Location	Length of Defect	Remarks	SKETCH

14

Liquid Penetrant Testing for Pressure Vessels

1 SCOPE

The test procedure is written in accordance with ASME Section V Article 6, Liquid Penetrant Examination for pressure vessels fabricated under ASME Section VIII Div (1).

2 QUALIFICATION OF PERSONNEL

2.1 Personnel performing the liquid penetrant examination shall be qualified in accordance with the company's written practice established in compliance with ASNT Recommended Practice SNT-TC-1 A (Personnel Qualification and Certification in Nondestructive Testing).

2.2 PT Level II or Level III certified personnel shall carry out liquid penetrant examination and evaluate the acceptability of any penetrant indication.

2.3 PT Level I certified personnel may carry out liquid penetrant examination under the supervision of a PT Level II personnel and shall follow approved procedures.

3 TECHNIQUE

Either a color contrast penetrant or fluorescent penetrant technique may be used. In both the cases the penetrant shall be removable by solvents.

4 SURFACE PREPARATION

In general, satisfactory results may be expected when the surface of the part is in the as-welded, as-rolled, as-cast, or as-forged condition. Surface preparation by grinding, machining, or other methods may be necessary where the surface irregularities could mask the indications of unacceptable discontinuities.

Prior to any liquid penetrant examination, the surface to be examined and all adjacent areas within at least 1 in. shall be dry and free of all dirt, grease, lint, scale, welding flux, weld spatter, oil, and other extraneous matter that could obscure surface openings or otherwise interfere with the examination. Cleaning agents that are to be used shall either be detergents or organic solvents.

5 DRYING AFTER PREPARATION

After cleaning, the drying of the surfaces to be examined shall be accomplished by normal evaporation or with forced hot air as appropriate. A minimum period of a 15-min lapse shall be permitted to ensure that the cleaning solution has evaporated prior to the application of the penetrant.

6 TESTING MEDIA

A solvent-removable system consisting of the following (aerosol spray type) will be used for the examination:

Magnaflux Cleaner SKC-NF-ZC 7
Magnaflux Penetrant SKL-HF/S
Magnaflux Developer ZP-9E

7 EXAMINATION

7.1 Technique for Standard Temperature

As a standard practice, neither the temperature of the penetrant nor the surface of the part to be examined shall be below 15.5°C (60°F) or above 51.6°C (125°F) throughout the examination period. Local heating or cooling is permitted, provided the part temperature remains in the range of 15.5°C (60°F) to 51.6°C (125°F) during the examination. Where it is not practical to comply with these temperature limitations, the test may be carried out at another temperature, provided the procedure adopted is qualified as specified in T-653 (Technique for Non-Standard Temperatures).

7.2 Penetrant Application

The penetrant is applied by spraying to the entire surface to be examined. The excess penetrant is allowed to drain from the part after application, while allowing for proper dwell time. In general, a dwell time of 5 to 10 min is sufficient for most of the applications.

7.3 Excess Penetrant Removal

After a dwelling time of 5 to 10 min, excess penetrant can be wiped off by a clean lint-free cloth, repeating the operation until most traces of penetrant have been removed. The remaining traces shall be removed by lightly wiping the surface with cloth or absorbent paper moistened with solvent. The direct spraying of the solvent onto the surface following the application of penetrant prior to developing is prohibited.

7.4 Drying After Excess Penetrant Removal

For the solvent-removal method, the surface may be dried by normal evaporation.

7.5 Developing

The developer shall be applied as soon as possible after the removal of the penetrant. The time interval shall not exceed 5 min under any circumstances. A nonaqueous developer shall be used with either color contrast or fluorescent penetrant.

8 INTERPRETATION

The true size and type of discontinuities are difficult to evaluate if the penetrant diffuses excessively into the developer. Consequently, the surface shall be closely observed during the application of the developer to monitor the behavior of indications, which tend to bleed out profusely. The final interpretation shall be made after allowing the penetrant to bleed out for 7 to 30 min. If bleed-out does not alter the examination results, a longer period is permitted. If the surface to be examined is large enough to preclude the complete examination within the prescribed time, the surface shall be examined progressively.

8.1 Color Contrast Penetrants

With color contrast penetrants, the developer forms a reasonably uniform white coating over the entire surface. Surface discontinuities are indicated as bleed-out of the penetrant, which is normally deep red in color, that stains the developer.

Indications in light pink may indicate excessive cleaning. Inadequate cleaning may leave an excessive background, making the interpretation very difficult. Adequate illumination is required to ensure sensitivity during the examination and evaluation of indications.

8.2 Fluorescent Penetrant

With fluorescent penetrants, the process is essentially the same as for color penetrants with the exception that the examination is performed in ultraviolet light, called black light. The examination shall be performed as follows.

1. It shall be performed in a darkened area.
2. The examiner shall be in the darkened area for at least 5 min prior to performing the examination to adapt to the darkness. If the examiner wears glasses or lenses, they shall not be photosensitive.
3. The black light shall be allowed to warm up for a minimum of 5 min prior to use or measurement taken of the intensity of the ultraviolet light emitted.
4. The black light intensity shall be measured with a black light meter. A minimum of 800 μ W/cm^2 on the surface of the part being examined is required. The black light intensity shall be measured at least once every 8 hr and whenever the workstation is changed.

9 ACCEPTANCE CRITERIA

The acceptance criteria of a component shall be evaluated in terms of the acceptance standards of ASME Section VIII Div (1) requirements.

10 POST-TEST CLEANING

Post-test cleaning is necessary in cases where residual penetrant or developer could interfere with subsequent processing or with service requirements. A simple water rinse, a machine wash, vapor degassing, or a solvent soak may be employed for the purpose.

11 REPORTING

All tests carried out shall be reported in a format that contains the following information:

1. Name of the client
2. Type of chemicals used
3. Identification of components tested
4. Method of test
5. Result of interpretation
6. Name and signature of the tester
7. Date of test.

12 SOLVENT REMOVABLE PENETRANT INSPECTION PROCESS

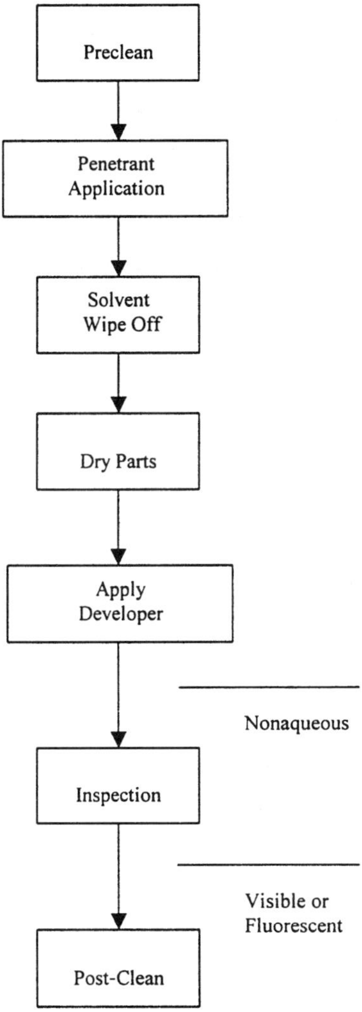

13 REPORT FORMAT

EMBLEM

LIQUID PENETRANT EXAMINATION REPORT

Order No.		Sheet	of
Purchaser		Attachment to Certificate No	
Description		Date	
Drawing No(s)			

Code Description of Surface Temperature of Surface Surface Condition Pre-Cleaning Penetrant (Make) Penetrant Application Dwell Time Cleaner (Make) Removal of Penetrant Developer (Make) Developer Application Developing Time Results Post-cleaning	
Remarks	

15

Furnace Postweld Heat Treatment for Pressure Vessels

1 SCOPE

This procedure is for the furnace postweld heat treatment (PWHT), or stress relieving, of fabricated pressure vessels manufactured from P. No. 1, 3, and/or 4 materials in accordance with ASME Section VIII Div (1).

2 STANDARDS

This specification is in line with the requirements of UCS 56 of ASME Section VIII Div (1), which is applicable to most of the carbon steel materials.

3 MATERIALS

This specification covers P. Nos. 1, 3, and 4 materials as classified in QW 422 of ASME Section IX.

4 EQUIPMENT

The furnace shall be capable of fulfilling the requirements incorporated in this specification. The heating method can be electrical or oil fired.

5 LOADING IN THE FURNACE

The following requirements shall be met with regard to loading the furnace:

1. The vessel shall be loaded in to the furnace at a temperature not exceeding 427°C (800°F).
2. The vessel shall be supported on suitable spacers and sections to a minimum height of 100 mm above the furnace hearth and in such a way as to permit free circulation of hot air as well as to prevent distortion during the stress relieving cycle. When items are to be stacked one above the other, they shall also be separated in the same fashion by providing suitable temporary structures. A clearance of a minimum 300 mm may be provided all around the walls of the furnace between the charge and the walls.
3. In oil or gas fired furnaces, there shall be no direct flame impingement onto the charge and there shall be no impediments to the free circulation of hot gases.

4. The serration on flange faces or specially machined and exposed surfaces shall be suitably protected against excessive scaling with clay or other approved compounds during stress relieving treatment.

6 TEMPERATURE MEASUREMENT

A sufficient number of thermocouples (at least four) shall be placed in intimate metallic contact with the vessel in such a way that they will record the actual metal temperature and give an indication of the maximum temperature differential within the vessel at different locations. Typical positions likely to show a variation in temperature from the average temperature are near to the door, the hearth, the roof, the flame posts, the heaviest and lightest sections of the vessel, etc. A typical placement of thermocouples is shown in the sketch.

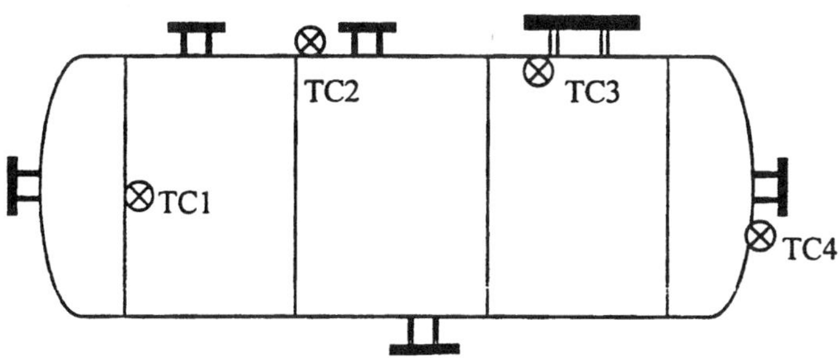

All thermocouples shall be checked before each application against any mechanical damages to confirm its fitness for use. They shall also be identified with the calibration records, which shall be recorded and maintained so that the records are available for verification by third parties.

Records shall also be checked before the actual conducting of the stress relief treatment. Spare recorders shall be available as a safeguard against failure of any recorder while in use.

All thermocouples and recorders shall be calibrated every 6 months or when malfunctioning is suspected. Records to this effect shall be maintained and shall be verified by the third-party inspection engineer.

7 METHOD OF ATTACHMENT OF THERMOCOUPLES

The method chosen shall be capable of indicating the true metal surface temperature and at the same time shall be free from the direct radiation or conduction from the heating source.

By means of a slotted nut or a bridge piece welded to the surface in the selected position, the thermocouples can be clamped under a set screw in contact with the surface of the component or weld. The nuts or bridge piece shall be of the smallest size possible and shall be made from compatible materials and tack welded to the vessel using a compatible electrode as specified in the applicable specification. A qualified welder per the applicable welding procedure specifications with the required preheat shall make the tack welds, and the size of the tack shall be the smallest possible. A slipper plate shall be provided between the thermocouple and the set screw so that mechanical damage to the thermocouple during clamping can be avoided.

8 HEAT TREATMENT

The heating rate, soaking temperature and time, temperature differential during soaking, and cooling rate shall be as follows.

1. The maximum heating rate permitted above 427°C (800°F) (the maximum recommended furnace tem-

perature at the time of loading the vessel in the furnace) is 204°C/hr per 25 mm (400°F/hr per inch) of section thickness subjected to a maximum of 204°C/hr (400° F/hr). The variation in temperature on the vessel during heating shall be less than 121°C (250°F) in any 4.57-m (15-ft) interval of length.

2. The soaking temperature shall be 593°C (1100°F) minimum. The maximum difference between the temperatures shown by different thermocouples on the vessel shall be 65.6°C (150°F). The soaking time shall be as follows:

Thickness of the vessel (mm)	P. No. 1 and 3, Gr. Nos. 1, 2, and 3	P. No. 4, Gr. Nos. 1 and 2
Up to 50	1 hr/25 mm with a minimum of 15 min	1 hr/25 mm with a minimum of 15 min
50–125	2 hr + 15 min for each additional inch over 2 in.	1 hr/25 mm
Over 125	2 hr + 15 min for each additional inch over 2 in.	5 hr + 15 min for each additional inch over 5 in.

3. The maximum cooling rate up to 427°C (800°F) shall be 260°C/hr per 25 mm (500°F/hr per inch) section thickness subjected to a maximum of 260°C/hr (500°F/hr). The cooling below 427°C (800°F) can be in still air. In any case, the rate of heating and cooling need not be less than 40°C/hr.
4. The variation in temperature on the vessel during heating shall be less than 93°C (200°F) in any 4.57-m (15-ft) interval of length of the vessel.

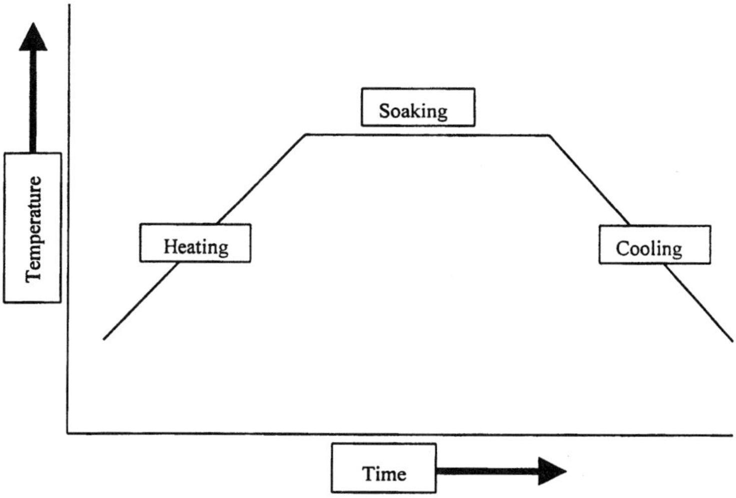

9 INSPECTION

The QAC department shall ensure the compliance to the various requirements of this procedure by verifying and clearing the following stages:

QAC check points	Checklist
Clearance for S/R	Ensure that all previous operations have been carried out satisfactorily.
Furnace loading	Charge details. Calibration of thermocouples and recorders. Loading and thermocouple attachment as per Sections 5 and 6.
After S/R	Prepare records as per Section 10. Check compliance and approve. In case of deviations from specifications, obtain decision from design.

10 HEAT TREATMENT RECORDS

Each stress relieving shall be documented in the format presented and shall contain the following essential information:

1. In order to distinguish the recording of each of the thermocouples, each recording pen shall be of a different color in the graph to plot the actual chart.
2. Full particulars of the charge, preferably with a sketch indicating the location of each component and identifying of the thermocouple fixed on it.
3. The temperature recorder chart shall also have the details of the charge typed on it with a cross-reference to the confirmation certificate.
4. The chart will also contain information like the chart speed, start and finish time, dates, etc.

11 REPORT FORMAT

EMBLEM

POSTWELD HEAT TREATMENT CONFIRMATION REPORT

Order No.		Sheet	of
Purchaser		Attachment to Certificate No	
Description		Date	
Drawing No(s)			

Description of items Post-Weld Heat Treated	
Material Specification	

DETAILS OF POSTWELD HEAT TREATMENT

PWHT Chart Number	
PWHT Method	
Type of Heating	
Heating Rate	
Required	
Actual	
Soaking Temperature &Time	
Required	
Actual	
Cooling Rate	
Required	
Actual	
Thermocouple Type	
Recorder Identification	
Recorder Chart Speed	
Colour of Recorder Points	
Recorder Calibration Details	
Hardness test Results	
Type	
Required	
Actual	
Remarks	

16

Hydrostatic Testing of Pressure Vessels

1 SCOPE

The procedure is prepared based on the requirements specified in ASME Section VIII Div (1) as well as on good engineering practices followed by renowned consultants.

All shop fabricated pressure vessels shall undergo a final hydrostatic test at the pressure mentioned in the drawings upon the satisfactory completion of the following stages of fabrication.

1. All fabrication activities have been completed except for those operations which could not be performed prior to the test such as weld end preparation and

cosmetic grinding on the base material which does not affect the required thickness.

2. All examinations and tests have been performed with satisfactory results except those required after the hydro test.

2 PREPARATION FOR THE TEST

Horizontal vessels resting on saddles can be hydro tested on the same saddle. However, precautions have to be taken to arrest any overturning movements of the vessel during testing, as there is no anchoring provided. If the saddle is too short to facilitate the inspection of the weld joints at the underside of the vessel, sufficient packing may be provided below the saddles to have free access for inspection.

In the case of long vertical columns, it is not possible to carry out the test in that position due to safety considerations. Therefore, it is always advisable to carry out the test in the horizontal position. In the absence of supports for holding the vessel in that position, temporary saddles may be provided at required locations based on the hydrostatic test weight and the rigidity of the vessel. The vessel may be arrested properly by driving wedges between the saddle and vessel in the absence of any welding with this temporary saddle.

While positioning the vessel for test, care shall be taken to ensure that almost all the big openings are at the top of the vessel. This will facilitate filling up of the vessel as well as bleeding out the air entrapped in the vessel during filling.

All nozzles shall be blanked using flanges of required rating with good quality fasteners suitable for holding the required test pressure. As a good engineering practice, it is not recommended to use the same bolting material used for the hydro test during service. Therefore, hydrostatic testing should be carried out with a separate set of bolting which have to be supplied to the client for future in-service tests envisaged in the life-cycle of the vessel.

Suitable air vents shall be provided at the top points to facilitate the bleeding out of air prior to the pressurization.

Pressure gauges may be mounted at the topmost point on the vessel, which will indicate the lowest pressure in the vessel. The pressure gauges and the pressurization line to the vessel may be connected through isolation valves so that in the eventuality of failure of any of these items, the test can go on by replacing the defective items after the closure of the isolation valves.

A liquid relief valve set at $1\frac{1}{3}$ times the test pressure is recommended. While under test, the vessel is likely to be warmed up materially when personnel may be absent.

3 TEST MEDIUM

The medium of test shall be water, irrespective of the material of construction of the vessel. In the case of vessels made out of carbon steel or low alloy steel, the water used for testing shall contain rust inhibitors like 0.2% sodium nitrate. For stainless steel vessels, the maximum total chloride (ion) content permitted in the water shall be 20 ppm. In case this is not possible, 0.5% sodium nitrate solution shall be added to the water used for the hydrostatic test. However, this action needs the approval of the client or consultant prior to the testing.

4 PRESSURE GAUGES

A minimum of two pressure gauges shall be used for the hydrostatic test. These shall be mounted at the topmost point on the vessel and shall be placed after isolation valves so that in the event of malfunctioning of the gauge, it can be replaced without affecting the testing schedule.

The range of the gauges used shall be with in $1\frac{1}{2}$ to 4 times the proposed test pressure. Further the red line marked on the dial of the gauge (the reliable and linear range of the gauge) shall be slightly above the intended test pressure.

All gauges shall have valid calibration certificates on the date of test.

In the case of large diameter vessels, it is recommended to also use a pressure recorder to record the actual hydrostatic test. This recorder to a shall have valid calibration on the date of test. The range specified for the gauges will not be applicable in the case of recorder.

5 PRESSURE TEST

On filling up the vessel completely, the pressurisation pump is started with all the high point vents open. When the water starts bleeding freely through these vents, they are closed one by one while the pump is working.

When all the high point vents are closed, the pressure starts building up gradually, depending on the capacity of the pump. The required pressure shall be built up in steps of 25% of the test pressure, allowing for a settling time of 5 min after every step. In other words, the pressure shall be first built up gradually to 25% of the test pressure and to be held at that pressure for 5 min. Subsequently the pressure is raised to 50 and 75% of the test pressure in the same way, allowing a settling time of 5 min in between. As the inspection of all the weld joints is recommended at two-thirds of the test pressure, this inspection may be carried out at the 75% test pressure step.

In case every thing is satisfactory, the pressure may be raised to 100% of the test pressure and held at that pressure for at least half an hour. All joints may be checked again swiftly at this pressure for any leaks or failure.

After the satisfactory completion of the test, the pressure is released gradually to 50% of the test pressure and held at that level for about 5 min. Afterward, it is brought down to atmospheric pressure gradually, as done previously.

Subsequently the vessel is drained of all water and dried using hot air. In case of vessels with inside protective coating, the same is applied as per specification. Otherwise silica gel bags are suspended inside the vessel prior to the closure of the vessel.

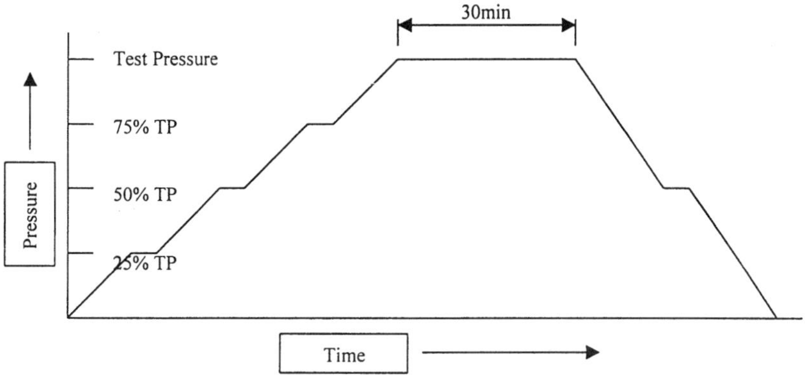

6 RECORDS

Records pertaining to the test shall be made in the format enclosed along with this procedure and shall be accompanied by all enclosures mentioned therein.

7 REPORT FORMAT

EMBLEM

HYDROSTATIC TEST REPORT

Order No.		Sheet	of
Purchaser		Attachment to Certificate No	
Description		Date	
Drawing No(s)			

Test Pressure Test temperature Test Medium Duration Pressure Gauge Range Pressure Gauge/ Recorder Identification Pressure Gauge Calibration Details Result Date of Test Attachments
Remarks

17

Quality Assurance and Control in Welding

1 WELD QUALITY

In any production weld, the term *weld quality* is relative, meaning that the application determines what is good and bad. The required quality of the proposed weld is achieved by the judicious application of quality assurance and quality control (QAC) techniques developed for the specific purpose.

2 QUALITY ASSURANCE

In quality assurance, the degree of quality required is ascertained based on application needs, a standard to be established based on service needs, and also consideration of the engineering performance and safety factors.

The five *P*s that assure quality are

1. *Process selection*. The process selected must be the most appropriate one for the job under the given conditions on all considerations.
2. *Preparation*. The joint configuration must be right and compatible with the welding process.
3. *Procedure*. To assure uniform results, the process must be clearly spelled out in detail and followed religiously during welding.
4. *Pretesting*. By a full-scale mock-up test, the process, preparation, and procedures are proved to have the desired standard of quality.
5. *Personnel*. Personnel assigned for the job shall be qualified to perform the assigned tasks.

3 QUALITY CONTROL

In quality control, the standards or norms set by the QA are followed without omission. Once this status is achieved, the weld will have the required quality within the tolerance permitted by the standard.

4 STANDARDS

Different countries have their own specification for the qualification of welding procedures and welders. The most commonly used ones in India for the manufacture of pressure vessels are

1. ASME Section IX
2. EN 287 and EN 288

Among these, most extensively used is ASME Section IX. This is the qualification standard for welding and brazing procedures, welders, brazers, and welding and brazing operators.

ASME Section IX is divided in to two basic sections, Part QW and QB, along with Appendices A, B, and D. Part QW deals with welding and Part QB deals with brazing. Appendix A is mandatory and deals with the submittal of technical enquiries to the boiler and pressure vessel committee. Appendix B is nonmandatory and gives sample forms for welding and brazing. Appendix D gives nonmandatory P. number listings. As we are concerned with welding, Part QW is elaborated on here. Part QW is further subdivided into four articles, the details of which are as follows:

Article I	**Welding: General Requirements**	
QW 100		General: explains the jurisdiction of the standard
	QW 101	Scope
	QW 102	Terms and definitions
	QW 103	Responsibility
QW 110		Weld orientation: illustrates various orientations of welds
QW 120		Test positions of groove welds
	QW 121	Plate positions
	QW 122	Pipe positions
	QW 123	Test positions for stud welds
QW 130		Test positions for fillet welds
	QW 131	Plate positions
	QW 132	Pipe positions
QW 140		Types and purposes of tests and examinations
	QW 141	Mechanical tests
	QW 142	Special examinations for welders
	QW 143	Examination for welding operators
	QW 144	Visual examination
QW 150		Tension tests
	QW 151	Specimens
	QW 152	Tension test procedures
	QW 153	Acceptance criteria—tension tests

QW 160		Guided bend tests
	QW 161	Specimens
	QW 162	Guided bend test procedure
	QW 163	Acceptance criteria
QW 170		Notch toughness tests
	QW 171	Notch toughness tests—charpy V notch
	QW 172	Notch toughness tests—drop weight
QW 180		Fillet weld tests
	QW 181	Procedure and performance qualification specimens
	QW 182	Fracture tests
	QW 183	Macroexamination—procedure specimen
	QW 184	Macroexamination—performance specimen
QW 190		Visual examination—performance
	QW 191	Radiographic examination
	QW 192	Stud weld tests—procedure qualification specimens
	QW 193	Stud weld tests—performance qualification specimens
	QW 195	Liquid penetrant examination
	QW 196	Resistance weld testing
Appendix 1		Rounded indications chart
Article II	**Welding Procedure Qualification**	
QW 200		General: explains the jurisdiction of WPS
	QW 201	Manufacturer's responsibility
	QW 202	Types of tests required
	QW 203	Limits of qualified positions for procedures
QW 210		Preparation of test coupons
	QW 211	Base and filler metal
	QW 212	Types and dimensions of groove welds
	QW 213	P. No. 11 base metals
	QW 214	Corrosion resistant weld metal overlay
	QW 215	Electron beam and laser beam welding
	QW 216	Hard facing weld metal overlay
	QW 217	Joining of composite (clad metals)
	QW 218	Applied lining
QW 250		Welding variables
	QW 251	General
	QW 252	Oxy fuel gas welding (OFW)
	QW 253	Shielded metal arc welding (SMAW)
	QW 254	Submerged arc welding (SAW)
	QW 255	Gas metal arc and flux cored arc welding (GMAW/FCAW)

	QW 256	Gas tungsten arc welding (GTAW)
	QW 257	Plasma arc welding (PAW)
	QW 258	Electro slag welding (ESW)
	QW 259	Electro gas welding
	QW 260	Electron beam welding
	QW 261	Stud welding
	QW 262	Inertia or continuous drive friction welding
	QW 263	Resistance welding
	QW 264	Laser beam welding
	QW 283	Welds with buttering
	QW 284	Resistance welding equipment qualification
	QW 285	Resistance spot and projection weld procedure qualifications
	QW 286	Resistance seam weld procedure qualification
Article III	**Welding Performance Qualification**	
QW 300		General
	QW 301	Tests
	QW 302	Types of tests required
	QW 303	Limits of qualified positions and parameters
	QW 304	Welders
	QW 305	Welding operators
	QW 306	Combination of welding processes
QW 310		Qualification test coupons
QW 320		Retest and renewal qualification
	QW 321	Retests
	QW 322	Expiration and renewal qualification
QW 350		Welding variables for welders
	QW 351	General
	QW 352	Oxy fuel gas welding
	QW 353	Shielded metal arc welding
	QW 354	Submerged arc welding
	QW 355	Gas metal arc and flux cored arc welding
	QW 356	Gas tungsten arc welding
	QW 357	Plasma arc welding
QW 360		Welding variables for welding operators
	QW 361	General
	QW 362	Electron beam, laser beam, and friction welding
	QW 363	Stud welding
QW 380		Special processes
	QW 381	Corrosion resistant weld metal overlay
	QW 382	Hard facing weld metal overlay
	QW 383	Clad plate or applied lining
	QW 384	Resistance welding operator qualification

Article IV	**Welding Data**	
QW 400		Variables
	QW 401	General
	QW 402	Joints
	QW 403	Base metals
	QW 404	Filler metals
	QW 405	Positions
	QW 406	Preheat
	QW 407	Postweld heat treatment
	QW 408	Gas
	QW 409	Electrical characteristics
	QW 410	Techniques
	QW 416	Welding variables (welding performance chart)
QW 420		Materials grouping
	QW 422	P and S numbers
	QW 423	Alternative base materials for welder qualification
	QW 424	Base materials used for procedure qualification
QW 430		F numbers
	QW 431	General
	QW 432	F numbers
	QW 433	Alternative F numbers for welder qualification
QW 440		Weld metal chemical composition
	QW 441	General
	QW 442	A numbers
QW 450		Specimens
	QW 451	Procedure qualification thickness limits and test specimens
	QW 452	Performance qualification thickness limits and test specimens
	QW 453	Procedure/performance qualification thickness limits and test specimens for hard facing and corrosion resistant weld metal overlay
QW 460		Graphics
	QW 461	Positions
	QW 462	Test specimens
	QW 463	Order of removal
	QW 466	Test jigs
	QW 469	Typical test joints
QW 470		Etching—process and reagents
	QW 471	General
	QW 472	For ferrous metals
	QW 473	For nonferrous metals
QW 490		Definitions
	QW 491	General
	QW 492	Definitions

5 WELDING PROCEDURE SPECIFICATION

The welding procedure specification (WPS) is a proposal for welding a joint between two specified base metals. The purpose of a WPS is to define and document the details that are to be carried out in specific materials or parts. To fulfill the purpose efficiently, the WPS should be as concise and clear as possible without extraneous details.

From a designer's point of view, the properties of the weld joint are designated as follows:

1. Yield stress
2. Ultimate tensile strength
3. Percentage elongation
4. Impact strength and so on

The welding procedure specifications are written exactly to these property requirements onto relevant welding parameters comprising of mainly two factors;

1. Physical soundness (free from discontinuities)
2. Metallurgical compatibility

Physical soundness is related to the mode of deposition. In other words, the process techniques are influenced by the skill of the welder or welding machine operator. Metallurgical compatibility is by itself dependent upon the chemical composition of the weld metal, filler metal, flux or gas, and the heat cycles the weldment undergoes during welding and postweld heat treatment.

Radiography can give conclusive account of physical soundness of the weld, whereas the metallurgical compatibility is revealed only by conducting a mechanical test.

ASME Section IX recommends formats for WPS in QW 482 (enclosed at the end of this chapter) wherein the following details as shown below shall be filled in:

1.	Joints	QW 402
2.	Base metals	QW 403
3.	Filler metals	QW 404
4.	Positions	QW 405
5.	Preheat	QW 406
6.	Post weld heat treatment	QW 407
7.	Gas	QW 408
8.	Electrical characteristics	QW 409
9.	Techniques	QW 410

The details are to be filled in depending on applicability to the particular welding process selected in the WPS.

6 BASIC INFORMATION FOR THE WPS

The information needed for the WPS includes

1. Base metals to be joined
2. Typical weld joint configuration
3. Dimensions of parts to be joined
4. Welding process to be employed

Of these, the first three details could be gathered from either the purchase order or the technical specification or from the applicable standards, whereas the fourth could be decided based on the manufacturer experience.

7 PROCEDURE QUALIFICATION RECORD

The five basic steps involved in the qualification of a welding procedure are

1. Preparation and welding of suitable samples
2. Testing of representative specimens
3. Evaluation of overall preparation, welding, testing, and end results

4. Possible changes in procedure
5. Approval

As per ASME Section IX, the test requirements for a procedure qualification record (PQR) are spelled out in QW 451, the test specimens in QW 462, and order of removal of samples in QW 463. The acceptance criteria for welding are specified in QW 150, QW 160, QW 170, QW 180, and QW 190. The suggested format for the preparation of PQR is given QW 483 (enclosed at the end of this chapter).

8 WELDING PERFORMANCE QUALIFICATION

Welder, welding operator, and tack welder qualification tests determine the ability of those tested to produce acceptably sound welds with processes, materials, and procedures called for in the test. The qualification tests are not intended to be used as a guide for welding during actual construction, but rather to assess whether an individual has a certain minimum level of skill. They do not tell what an individual normally will or can do in production. For this reason, complete reliance should not be placed on qualification testing of welders. The quality of all production welds should be determined by inspection during and following completion of actual welding.

ASME Section IX lists the welding variables applicable for performance in QW 350 and QW 360. Allowable visual, mechanical, and radiographic examination requirements are described in QW 304 and QW 305. Retests and renewal of qualifications are given in QW 320. The suggested format for performance qualification tests is given in QW 484 (enclosed at the end of this chapter).

9 WELDING VARIABLES

Depending on their influence on obtaining the desired quality weldments, the welding parameters are listed and classified as essential, supplementary essential, and nonessential variables. These welding variables are described in Article IV: Welding Data under clauses QW 400 to QW 410. These welding variables are obviously listed for all the welding processes addressed in ASME Section IX. Hence for a process under consideration, only a few among them may be applicable, and these are given in sections QW 252 to QW 265 for welding procedure and QW 352 to QW 357 for performance qualification, and QW 361 to QW 384 for welding machine operators, special processes, etc., against each welding process.

9.1 Essential Variables

Essential variables are those in which a change, as described in specific variables, is considered to affect the mechanical properties of the weldment and shall require requalification of the WPS.

9.2 Supplementary Essential Variables

Supplementary essential variables are those variables which are also to be considered as essential variables when other sections of the code call for notch toughness testing.

9.3 Nonessential Variables

Nonessential variables are those in which a change, as described in specific variables, may be made in the WPS with out requalification.

10 MATERIAL GROUPING

10.1 P Numbers

To reduce the number of welding and brazing procedure qualifications required, base metals have been assigned P numbers and, for ferrous metals which have specified impact test requirements, group numbers within P numbers. These assignments are based essentially on comparable base metal characteristics, such as composition, weldability, brazability, and mechanical properties. However, these assignments do not imply that the base metal used in the qualification test can be selected without the consideration of compatibility from the standpoint of metallurgical properties, postweld heat treatment, design, mechanical properties, and service requirements. Where notch toughness is a consideration, it is presupposed that the base metals meet the specific requirements.

10.2 S Numbers (Nonmandatory)

S numbers group materials which are acceptable for use by ASME B 31, Code for Pressure Piping, or by selected boiler and pressure vessel code cases, but which are not included in ASME Boiler and Pressure Vessel code material specification (ASME Section II). These materials are grouped in S numbers similar to the P numbers. These are given in section QW 422 of ASME Section IX.

10.3 F Numbers

The F number groupings of electrodes and welding rods in QW 432 are based essentially on their usability characteristics, which fundamentally determine the ability of the welder to make satisfactory welds with a given filler metal. This grouping is again made to reduce the number of welding procedure and performance qualifications, where this can be done logically. However, the indiscriminate substitution of base metals and filler metals without consideration of compatibility is not permitted.

10.4 A Numbers

Based on the weld metal analysis, types of weld deposits are grouped into twelve A numbers, as given in QW 442.

11 ENCLOSURES TO VARIOUS DOCUMENTS

Enclosures to WPS / PQR

1. WPR form filled completely
2. PQR form including test results
3. Base material test certificate
4. Electrode batch test certificate
5. PWHT chart and calibration report of recorders and thermocouples used
6. Nondestructive test reports
7. Mechanical test reports if done by an outside agency
8. Other test reports, if any

Enclosures to WQT

1. WQT form filled completely
2. Mechanical test reports, if carried out
3. Other test reports, if any

or

1. Radiography test report
2. Other test reports, if any

12 PRODUCTION WELD QUALITY

When the WPS proposed is established by a satisfactory PQR, the WPS is claimed to be of required quality for the specific application. The welder who does the welding for the PQR qualification test is also given an automatic qualification in the applicable range. Subsequently, a sufficient number of welders are qualified depending on the requirements to meet production targets.

The establishment of the WPS/PQR and the WQT is the first step of quality control in welding. However, the quality of the final production weld is very much dependent on the quantum of quality control checks imposed in the actual manufacturing process by means of inspection tools like destructive and nondestructive testing methods, as described earlier.

Of the following figures, QW 482, QW 483, and QW 484 are reproduced from ASME Boiler and Pressure Vessel Code, Section IX by permission of the American Society of Mechanical Engineers, all rights reserved.

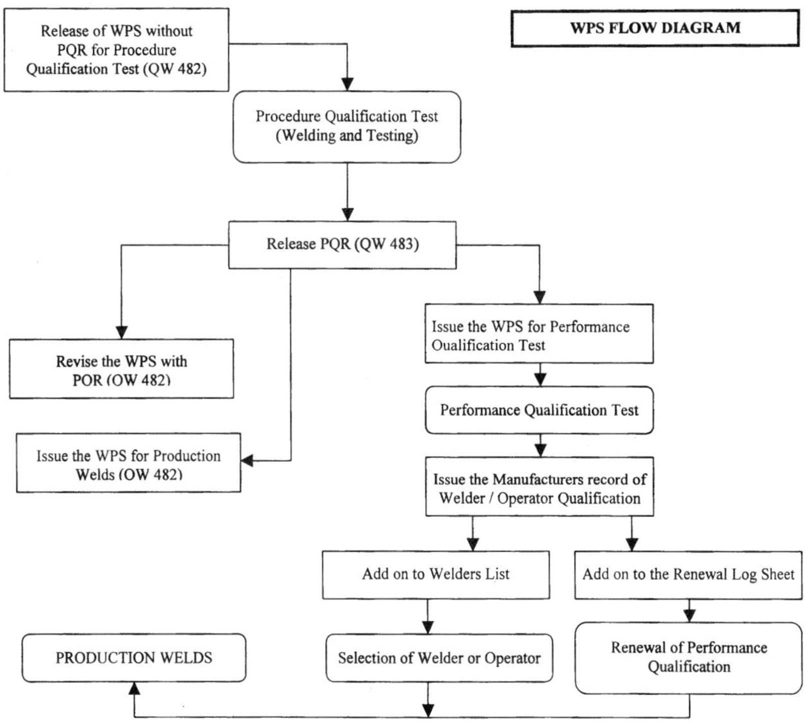

FLOW CHART FOR PREPARATION OF WELDING PROCEDURE SPECIFICATION AND PROCEDURE QUALIFICATION RECORD

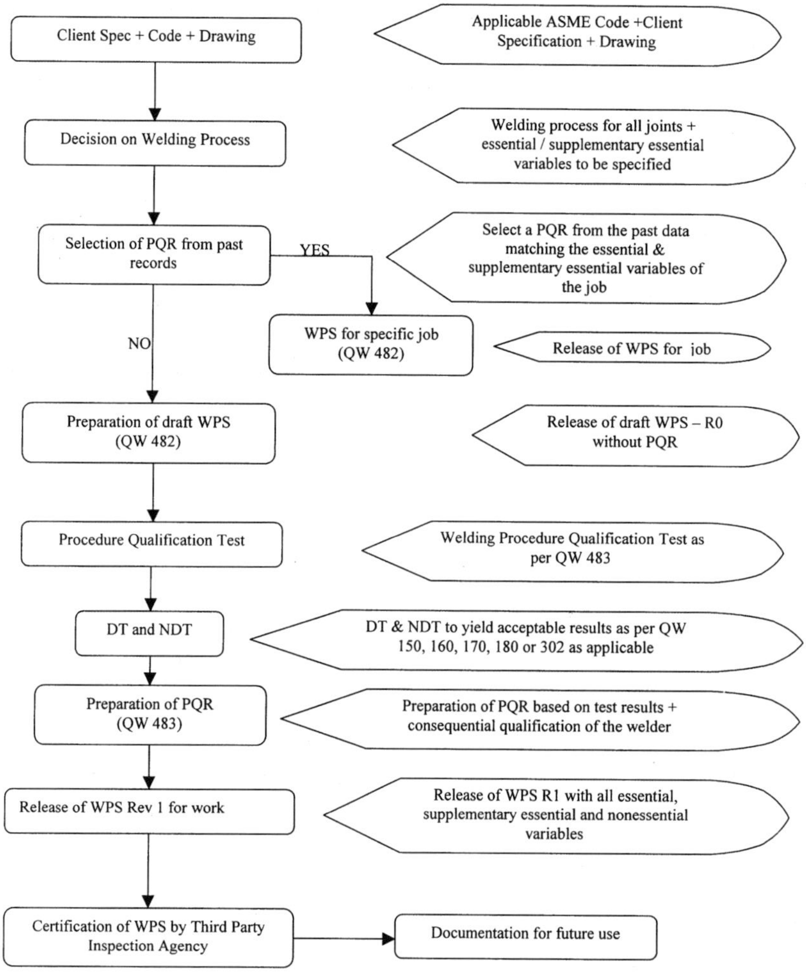

FLOW CHART FOR WELDER / WELDING OPERATOR QUALIFICATION RECORD

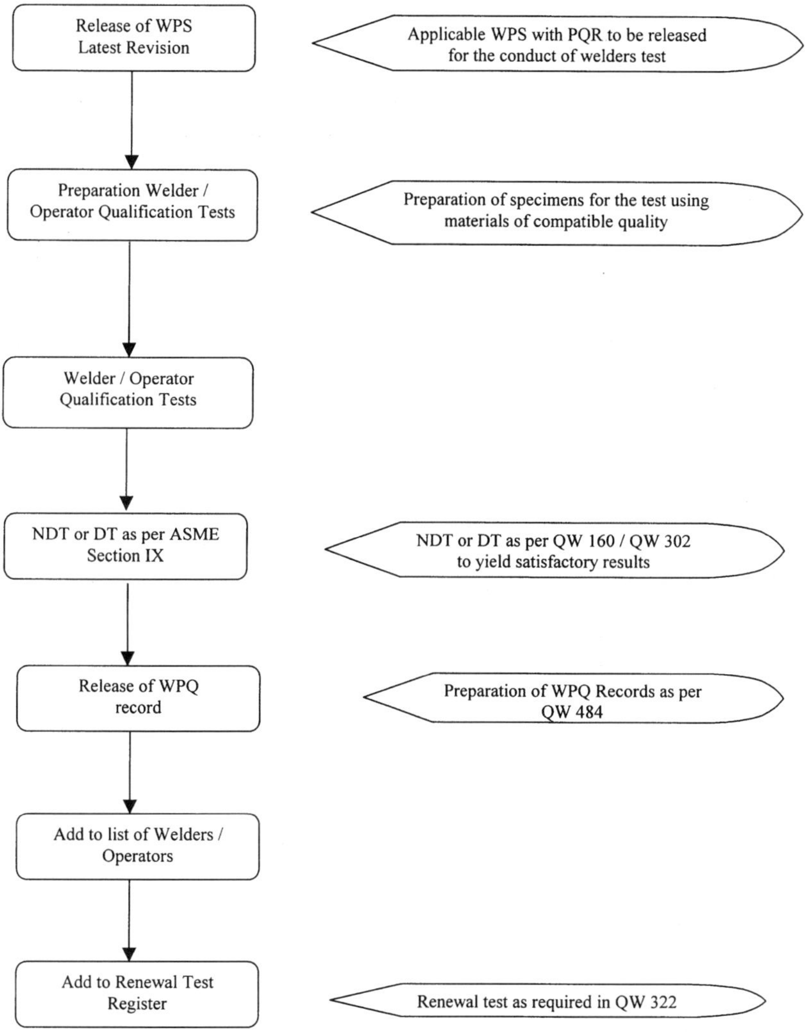

QW-482 FORMAT FOR WELDING PROCEDURE SPECIFICATION (WPS)
(See QW- 200.1, Section IX, ASME Boiler and Pressure Vessel Code)

Company Name By:
Welding Procedure Specification No. Date
Revision No. Date **Supporting PQR No.(s)**
Welding Process(es) Type(s)
(Automatic, Manual, Machine or Semi-Auto.)

JOINTS (QW-402)
Joint Design
Backing (Yes) (No)
Backing Material (Type)
(Refer to both backing and retainers)
☐ Metal ☐ Nonfusing Metal
☐ Nonmetallic ☐ Other
Sketches, Production Drawings, Weld Symbols or Written Description should show the general arrangement of the parts to be welded. Where applicable, the root spacing and the details of weld groove may be specified.

(At the option of the manufacturer, sketches may be attached to illustrate joint design, weld layers and bead sequence, e.g., for notch toughness procedures, for multiple process procedures, etc.)

BASE METALS (QW-403)
P-No. Group No. to P-No. Group No.
OR
Specification type and grade
To Specification type and grade
OR
Chem. Analysis and Mech. Prop.
To Chem. Analysis and Mech. Prop.
Thickness Range:
Base Metal: Groove Fillet
Pipe Dia. Range Groove Fillet
Other

FILLER METALS (QW-404)
Spec. No. (SFA)
AWS No. (Class)
F-No.
A-No.
Size of Filler Metals
Weld Metal
Thickness Range:
Groove
Fillet

Electrode Flux (Class)
Flux Trade Name
Consumable Insert
Other

POSITIONS (QW-405)	POSTWELD HEAT TREATMENT(QW-407)
Position(s) of Groove Welding Progression: Up Down Position(s) of Fillet	Temperature Range Time Range
PREHEAT (QW-406) Preheat Temp. Min. Interpass Temp. Max. Preheat Maintenance (Continuous or special heating where applicable should be recorded)	GAS (QW-408) Percent Composition Gas(es) (Mixture) Flow Rate Shielding Trailing Backing

ELECTRICAL CHARECTERISTICS (QW-409)
Current AC or DC Polarity
Amps (Range) Volts (Range)
(Amps and volts range should be recorded for each electrode size, position and thickness, etc. This information may be listed in a tabular form similar to that shown below.)
Tungsten Electrode Size and Type
(Pure Tungsten, 2% Thoriated, etc.)
Mode of Metal Transfer for GMAW
(Spray arc, short circuiting arc, etc.)
Electrode Wire feed speed range

TECHNIQUE (QW-410)
String or Weave Bead
Orifice or Gas Cup Size
Initial and Interpass Cleaning (Brushing, Grinding, etc.)

Method of Back Gouging
Oscillation
Contact Tube to Work Distance
Multiple or Single Pass (per side)
Multiple or Single Electrodes
Travel Speed (Range)
Peening
Other

Weld Layer(s)	Process	Filler Metal		Current		Volt Range	Travel Speed Range	Other / Remarks
		Class	Dia.	Type Polarity	Amp. Range			

QW-483 FORMAT FOR PROCEDURE QUALIFICATION RECORDS (PQR)

(See QW-200.2, Section IX, ASME Boiler and Pressure Vessel Code)

Company Name
Procedure Qualification Record No. Date
WPS No.
Welding Process(es)
Types (Manual, Automatic, Semi- Auto.)
JOINTS (QW-402)

Groove Design Test Coupon
(For combination qualifications, the deposited weld metal thickness shall be recorded for each filler metal or process used)

BASE METALS (QW-403) Material Spec. Type or Grade P-No. to P-No. Thickness of Test Coupon Diameter of Test Coupon Other	POSTWELD HEAT TREATMENT(QW-407) Temperature Time Other GAS (QW-408) Percent Composition Gas(es) (Mixture) Flow Rate Shielding Trailing Backing
FILLER METALS (QW-404) SFA Spec. AWS Class F-No. A-No. Size of Filler Metal Other Weld Metal Thickness	ELE. CHARACTERISTICS(QW 409) Current Polarity Amps. Volts Tungsten Electrode Size Other
POSITION (QW-405) Position of Groove Weld Progression (Uphill, Downhill) Other PREHEAT (QW-406) Preheat Temp. Interpass Temp. Other	TECHNIQUE (QW-410) Travel Speed String or Weave Bead Oscillation Multi Pass or Single Pass (per side) Single or Multiple Electrodes Other

PQR No.						
TENSILE TEST (QW-150)						
Specimen No.	Width mm	Thickness mm	Area mm^2	Ultimate Total Load kg	Ultimate Unit Stress kg/mm^2	Type of Failure and Location

GUIDED BEND TEST (QW-160)	
Type and Figure No.	Result

TOUGHNESS TEST (QW-170)							
Specimen No.	Notch Location	Specimen Size	Test Temp.	Impact Values			Drop Weight Break(Y/N)
				Kg-m	% Shear	Mils	

Comments

FILLET WELD TEST (QW-180)

Result – Satisfactory :Yes No Penetration into Parent Metal :Yes No

Macro – Results

OTHER TESTS

Type of Test

Deposit Analysis

Other

Welder's Name Clock No. Stamp No.

Tests conducted by: Laboratory Test No.

We certify that the statements in this record are correct and that the test welds were prepared, welded, and tested in accordance with the requirements of Section IX of the ASME Code.

Manufacturer

BY

Date

QW-484 FORMAT FOR WELDER/WELDING OPERATOR PERFORMANCE QUALIFICATIONS (WPQ)

(See Qw-301, Section IX, ASME Boiler and Pressure Vessel Code)

Welder's name Clock no. Stamp no.
Welding process(es) used Type
Identification of WPS followed by welder during welding of test coupon
Base material(s) welded Thickness

Manual or Semiautomatic Variables for Each Process(QW-350)	**Actual Values**	**Range Qualified**
Backing (metal, weld metal, welded from both sides, flux, etc) (QW-402)		
ASME P- No. to ASME P-No. (QW-403)		
< > Plate < > Pipe (enter diameter, if pipe)		
Filler metal specification (SFA) Classification (QW-404)		
Filler metal F-No.		
Filler metal variety for GTAW or PAW		
Consumable insert for GTAW or PAW		
Weld deposit thickness for each welding process		
Welding position (1G, 5G, etc.) (QW-405)		
Progression (uphill/downhill)		
Backing gas for GTAW, PAW, or GMAW; fuel gas for OFW (QW-408)		
GMAW transfer mode (QW-409)		
GTAW welding current type/polarity		

Machine Welding Variables for the Process used (QW-360)	**Actual Values**	**Range Qualified**
Direct/remote visual control		
Automatic voltage control (GTAW)		
Automatic joint tracking		
Welding position (1G, 5G, etc.)		
Consumable insert		
Backing (metal, weld metal, welded from sides, flux, etc.)		

Guided Bend Test Results

Guided Bend Tests Type < > QW-462.2 (Side) Results < > QW-462.3(a) (Trans. R & F) Type < > QW-462.3(b) (Long R & F) Results

Visual examination results (QW-302.4)
Radiographic test results (QW-304 and QW-305)
(For alternative qualification of groove welds by radiography)
Fillet Weld -- Fracture test Length and percent of defects in.
Macro test fusion Fillet leg size in. X in. Concavity/convexity in.
Welding test conducted by Laboratory test no.

We certify that the statements in this record are correct and that the test coupons were prepared, welded, and tested in accordance with the requirements of Section IX of the ASME Code.

Organization

Date By

18

Quality Plan

The following pages detail the quality plan in tabular form.

Owner: Document No.
Client: Rev. No.
Project/Location:

Legends
GM: General Manager
M(QAC): Manager (Quality Assurance & Control)
M(Prod): Manager (Production)
M(Des): Manager (Design & Drawing)
M(Com): Manager (Commercial)
M(Plg): Manager (Planning)
AM(Proc): Asst. Manager (Procurement)
Off(Store): Officer (Stores)
E(Maint): Engineer (Maintenance)

Item No.	Activity Element	Responsibility	Applicable document	Verifying document/action	Remarks
1	*Management responsibility.* Quality policy shall be specific to the functioning of the organization aimed at the achievement of corporate goals of the company and shall meet the expectations and needs of the customers.	Per organization chart	Quality assurance manual	Statement of authority & responsibility	
2	*Quality system.* A quality system shall be established and documented for ensuring that the product conforms to the customer requirements.	M(QAC)	Quality assurance manual	Audit report	
3	*Contract review.* A detailed review of the contract shall be carried out before the submission of tender to ensure the capability to meet the contract requirements and shall be documented.	M(Prod)	Quality assurance manual		
	A detailed comparison of the order with the submitted tender shall be carried out to consolidate the deviations from the original tender and to take appropriate action and shall be documented.	M(Com) M(Prod)			

4	*Design control*. Establish and document procedures to control and verify the design of the product in order to ensure that the specified requirements are met.	M(Des)	Quality assurance manual	Design calculations and drawings
5	*Document and data control*. Maintain and control the document and data related to the customer requirements with a master list or document control index identifying the status of current revision.	M(Com) M(Plg) M(QAC) M(Des) M(Prod)	Quality assurance manual	Quality manual and internal audit
6	*Purchasing*. Select the supplier on the basis of previous vendor qualification exercise as well as the past experience of their capability to deliver goods at best prices and in stipulated time frames and documentation pertaining to the same.	AM(Proc)	Review of PO and work order	Vendor evaluation record
7	*Control of customer supplied products*. Establish and maintain document for the control, verification, and storage of customer supplied products.	M(Prod) M(Plg) M(QAC) Off(Store)	Quality assurance manual	Material control record
8	*Product identification and traceability*. Establish standard methods and formats for recording the traceability of materials in process up to the final product as well as on the remaining raw materials to be used later.	M(Prod) M(QAC) Off(Store)	Quality assurance manual	Traceability records

Item No.	Activity Element	Responsibility	Applicable document	Verifying document/action	Remarks
9	*Process control.* Ensure the use of suitable equipment and workforce in the appropriate scope.	M(Plg) M(Prod) E(Maint)	Quality assurance manual, applicable WPS, and client specifications	PQR, welding records, and reports	
	Ensure the compliance with applicable codes and standards, quality plan and the documented procedures referenced.	M(Prod) M(QAC)			
	Monitor and control for suitable process parameters and product characteristics.	M(Prod) M(QAC)			
10	*Inspection and testing.* Ensure that the inspection and testing activities are carried out to verify the compliance of specified requirements for the product.	M(QAC) M(Prod)	Quality assurance manual	Inspection & test plan, inspection reports, and test reports	
	Receiving inspection and testing shall be carried out to ensure that the incoming materials comply with the customer requirements.	M(plg) Off(Store) M(QAC)			
	In-process inspection and testing shall be carried out according to the inspection and test plan to ensure the final quality of the product as envisaged in the contract.	M(QAC)			
	Maintain records as to all the inspection and tests carried out for future verification.	M(QAC)			

11	*Control of inspection, measuring, and test equipment.* Ensure the calibration of all inspection, measuring, and test equipment and maintain records as to the storage and handling of the equipment.	M(QAC)	Quality assurance manual	Calibration reports and internal audit
12	*Inspection and test status.* The inspection and test status shall be maintained as defined in the procedures as well as the ITP to ensure that only quality products are released for dispatch.	M(QAC)	Quality assurance manual	Inspection and test plan and internal audit
13	*Control of nonconforming products.* Product which does not conform to the specific requirements is prevented from progressing to the next stage of production. Nonconforming products shall be reworked to meet the specified requirements, accepted with or without repair or by concession, regraded for alternative use or rejected/scrapped. Repaired or reworked product shall be reinspected in accordance with inspection and test plan.	M(Des) M(Prod) M(QAC)	Quality assurance manual	NCR
14	*Corrective and preventive action.* Remedial action is to be taken to avoid nonconformities in future.	M(QAC) M(Prod) M(Des)	Quality assurance manual	Corrective action, inspection report, check on updated document, and review

Item No.	Activity Element	Responsibility	Applicable document	Verifying document/action	Remarks
15	*Handling, storage, packing, preservation and delivery*. The methods used for handling while in process as well as for the end product shall be safe and free from risk of damage. The method of storage shall be such that deterioration of the end product shall be avoided. Packaging and packing shall be in accordance with customer requirements. Preserve and segregate product as necessary when the product is under manufacturing control. On completion of the final testing and inspection, the product shall be delivered with protection as desired by the client.	M(Prod) M(QAC) AM(Proc) Off(Store)	Quality assurance manual	Inspect storage area, witness during shipping, and review of packing list	
16	*Control of quality records*. All quality records shall be legible and be retained in such a way that they are readily available for verification and shall be preserved without any deterioration.	M(QAC)	Quality assurance manual	Review of manufacturer's data report	
17	*Internal quality audits*. Internal quality audits shall be carried out to verify that the quality activities are followed religiously and meticulously for the effectiveness of the quality system. Findings of the quality audit shall be reported to the General Manager and the concerned departments to assess whether there is any possibility for improvement.	GM M(QAC)	Quality assurance manual	Audit report	

19

Inspection and Test Plan

The following pages detail the inspection and test plan in tabular form.

Owner:	Document No.
Client:	Revision No.
Project/Location:	Drawing No(s).
Equipment Name & No.	

Sl. No.	Activity element	Verifying document	Action dept.	Inspection/Tests			Remarks
				QAC	TP/ statutory	Client	
1	Incoming raw material inspection (heads, shell plates, pipes, flanges, forgings, fasteners, gaskets, and other pressure parts)	PO, code specifications	Stores Plg Proc	H	V	H	
2	Identification of heads and edge preparation, center hole plugging, etc.	Drawing, code	Prod	H	V	H	
3	Identification of shell plates, transfer of identification and marking on plates	Drawing, cutting plan	Prod	H	RW	H	
4	Bending of plates	Drawing, code	Prod	—	—	—	
5	Fit-up of longitudinal seam for welding	Drawing, code, WPS	Prod	H	RW	H	
6	Welding of longitudinal seam	Drawing, code, WPS	Prod	H	RW	H	
7	NDT of longitudinal seams	Drawing, code, NDT procedure	QAC QAC QAC	RT V UT W/V MPT&LPT RW/V	V RW/V V	V RW/V V	
8	Fit-up of shell-to-head, shell-to-shell, and shell-to-cone circumferential seams	Drawing, code, WPS	Prod	H	—	H	
9	Welding of circumferential seams	Drawing, code, WPS	Prod	H	RW	H	
10	NDT of circumferential seams	Drawing, code, NDT procedure	RT QAC UT QAC MPT&LPT QAC	V W/V RW/V	V RW/V V	V W/V RW/V	

11	Marking of nozzle orientation/ location and those of other attachments	Drawing, code	Prod	H	RW	H
12	Opening the shell for nozzles and manholes	Drawing, code	Prod	H	RW	H
13	Manufacture of subassemblies like manholes and nozzles	Drawing, code, WPS	Prod	H	RW	H
14	Fit-up of manholes and nozzles on the shell/head/cone	Drawing, code	Prod	H	W	H
15	Welding manholes and nozzles	Drawing code, WPS	Prod	H	RW	H
16	NDT of manhole and nozzle welds	Drawing, code, NDT procedure	UT QAC MPT&LPT QAC	W/V RW/V	V V	W/V RW/V
17	Fit-up and welding of other attachments like ladder, platform, and pipe supports, insulation clips, etc.	Drawing, code	Prod	H	—	H
18	Final dimensional check	Drawing, code	Prod QAC	H	V	H
19	Final visual inspection	Drawing, code	Prod QAC	H	W	H
20	Postweld heat treatment	Drawing, code, PWHT procedure	Prod	H/V	V	H/V
21	Re-NDT after postweld heat treatment	Drawing, code, NDT procedure	QAC	V	V	V
22	Final hydrostatic test	Drawing, code, HT procedure	Prod	H	H	H
23	Manufacture's data report	Client specification	QAC	H	V	H

H, hold point; V, verification of document; W, witness; RW, random witness.

Notes: (1) The heads, shell plates, nozzles, flanges, fastners, gaskets, etc., are considered to be bought out items as per the customer specifications reproduced in the purchase order; (2) Considered that the client inspector is permanently posted at the shop; (3) Third party/statutory inspection are envisaged only at certain important stages and may be modified based on specific requirements; (4) If needed by client, even more specific ITP can be prepared for each vessel by listing each and every seam, nozzle, etc., which will be essential in the case of very critical and, especially, very high pressure vessels.

20

Standard Tolerances for Pressure Vessels

The tolerances specified are in addition to the tolerances specified in applicable standards for the vessel.

1. The tolerance on diameter, considered as a difference between the nominal diameter and the actual diameter calculated based on the actual measured outside circumference divided by π, shall be within that indicated in the following table:

Sl. no.	Internal diameter (mm)	Tolerance (mm)
1	Up to 1200	± 3
2	1201–2000	± 5
3	2001–5000	± 7
4	5001 and above	± 8

2. For vessels with internal parts such as trays, grids, agitators, etc., the ovality measured as the difference between the greatest and the smallest diameter shall be less than or equal to 1% of the nominal diameter, with a maximum of 25 mm.
3. Tolerance on out-of-roundness in the case of vessels with external pressure shall be per the code requirement.
4. Tolerance on distance between the tangent lines shall be ±1.5 mm per meter, with a maximum of ±25 mm.
5. Tolerance on height from baseline for vertical vessels shall be ±3 mm for every 8 m of height, with a maximum of ±10 mm.
6. Tolerance on the height of nozzles from the axis of the vessel shall be ±3 mm.
7. Maximum permitted flange face out from the desired position shall be ±1/2°. For nozzles where agitators are connected, the tolerance is limited to 1/6°.
8. Maximum rotation of flange holes from the desired level: 1.5 mm
9. Tolerance on height of manhole flange from the vessel axis: ±6 mm.
10. Permitted face out of manhole flange in any direction: ±1°.
11. Tolerance on distance of flanged nozzle from reference line: ±5 mm for distance less than or equal to 3 m. Above 3 m, allowance may be calculated as ±1 mm per meter distance, with a maximum of ±25 mm.
12. Tolerance on manhole axis from the reference line: ±12 mm.
13. Tolerance on center to center distance of level indicator nozzles: ±1.5 mm.
14. Tolerance on distance between bottom discharge axis and reference line: ±3 mm.
15. Tolerance on nozzle orientation measured along the

circumference of the vessel, either between nozzles or from a reference point: ±3 mm for a distance less than or equal to 3 m and ±1 mm per meter for distances greater than 3 m, with a maximum of ±6 mm.

16. Tolerance on anchor bolt pitch circle diameter: ±3 mm for diameter up to 2000 mm and ±6 mm for diameters above 2000 mm.
17. Maximum deviation from horizontal plain of bearing surface of base ring or base plate:

Sl. no.	Internal diameter (mm)	Tolerance (mm)
1	Up to 1200	±3
2	1201–2000	±5
3	2001 and above	±7

18. Tolerance on height of anchor chair: ±6 mm.
19. Tolerance on positioning of anchor bolts with regard to main axis: ±6 mm (with a tolerance of distance between bolts of ±5 mm).
20. Tolerance on distance between reference line and base surface: +0 mm −10 mm.
21. Tolerance on distance between reference line and lugs: +0 mm −10 mm.
22. Tolerance on distance between tray support rings and reference line: ±6 mm.
23. Tolerance on distance between lower edge of downcomer and upper surface of the tray support ring: ±3 mm.
24. Tolerance on distance between support rings of adjacent trays: ±3 mm.
25. Tolerance on height of downcomer support: ±3 mm.
26. Tolerance on distance between vessel axis and internal face of downcomer support: ±3 mm.

27. Tolerance on the flatness of tray support rings: 0.8 mm for every chord of 300 mm. The maximum deviation of support surface from the horizontal, measured along the wall of the vessel in two points diametrically opposite to the tray, shall be 1.5 mm for vessels with internal diameter less than or equal to 1200 mm and 3 mm for vessels with internal diameter greater than 1200 mm.
28. Tolerance on distance between support clips or external welded rings and reference line or vessel axis: ±6 mm.
29. Tolerance on distance between external supports welded in pair: ±3 mm.

21

ASME Exhibits

This chapter consists of exhibits reproduced from ASME Boiler and Pressure Vessel Code Section VIII Div (1), by permission of the American Society of Mechanical Engineers.

1 UW 34: SPIN HOLES

Refer to Chapter 3, Section 5.3.

Spin holes are permitted at the center of heads to facilitate forming. Spin holes not greater in diameter than $2^3/_8$ in. may be closed with a full penetration weld using either a welded plug or weld metal. The weld and plug shall be no thinner than the head material adjacent to the spin hole.

The finished weld shall be examined and shall meet the acceptance requirements of Appendix 6 or Appendix 8 of this Division. Radiographic examination, if required by UW 11(a), and additional inspections, if required by the material specification, shall be performed.

This weld is a butt weld, but it is not categorized. It shall not be considered in establishing the joint efficiency of any part of the head or of the head-to-shell weld.

2 UG 81: TOLERANCE FOR FORMED HEADS

Refer to Chapter 3, Section 7.1.

(a) The inner surface of a torispherical, toriconical, hemispherical, or ellipsoidal head shall not deviate outside of the specified shape by more than 1¼% of *D* nor inside the specified shape by more than ⅝% of *D*, where *D* is the nominal inside diameter of the vessel shell at point of attachment. Such deviations shall be measured perpendicular to the specified shape and shall not be abrupt. The knuckle radius shall not be less than that specified.

(b) Hemispherical heads or any spherical portion of a torispherical or ellipsoidal head designed for external pressure shall, in addition to satisfying (a) above, meet the tolerances specified for spheres in UG 80(b) using a value of 0.5 for *L*/*Do*.

(c) Measurements for determining the deviations specified in (a) above shall be taken from the surface of the base metal and not from welds.

(d) The skirts of heads shall he sufficiently true to round so that the difference between the maximum and minimum inside diameters shall not exceed 1% of the nominal diameter.

(e) When the skirt of any unstayed formed head is machined to make a driving fit into or over a shell, the thickness shall not be reduced to less than 90% of that required for a blank head (see UW 13) or the thickness of the shell at the point of attachment. When so machined, the transition from the machined thickness to the original thickness of the head shall not be abrupt but shall be tapered for a distance of at least three times the difference between the thicknesses.

3 UG 80.1: PERMISSIBLE DEVIATION FROM CIRCULAR FORM FOR VESSELS UNDER EXTERNAL PRESSURE

Refer to Chapter 3, Section 7.2.

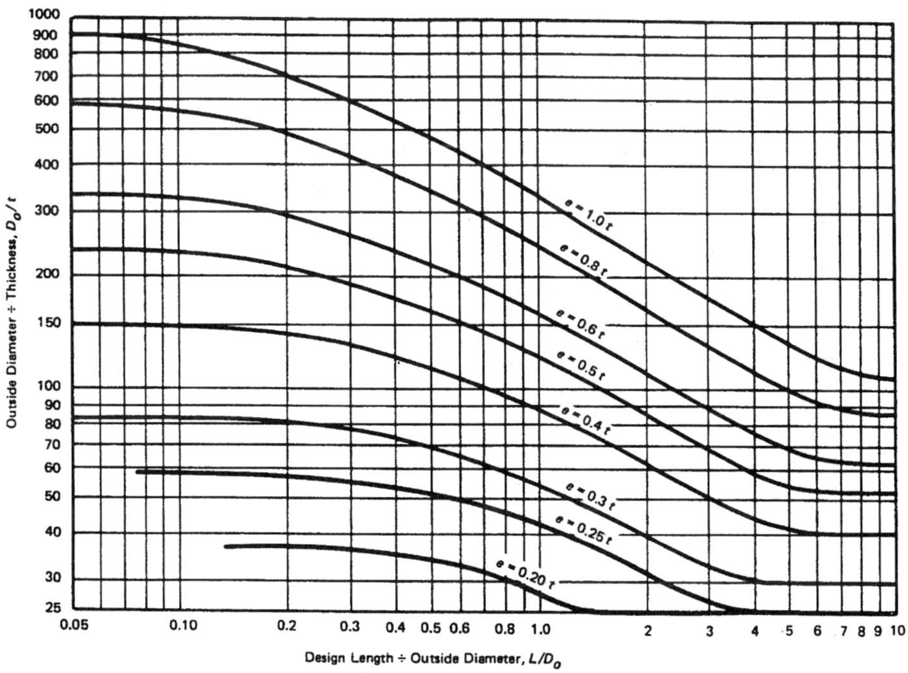

Figure 21.1 (ASME Figure UG 80.1) Maximum permissible deviation from a circular form e for vessels under external pressure.

4 UG 29.2: MAXIMUM UNSUPPORTED SHELL ARC FOR VESSELS UNDER EXTERNAL PRESSURE

Refer to Chapter 3, Section 7.2 and Chapter 4, Section 5.3.

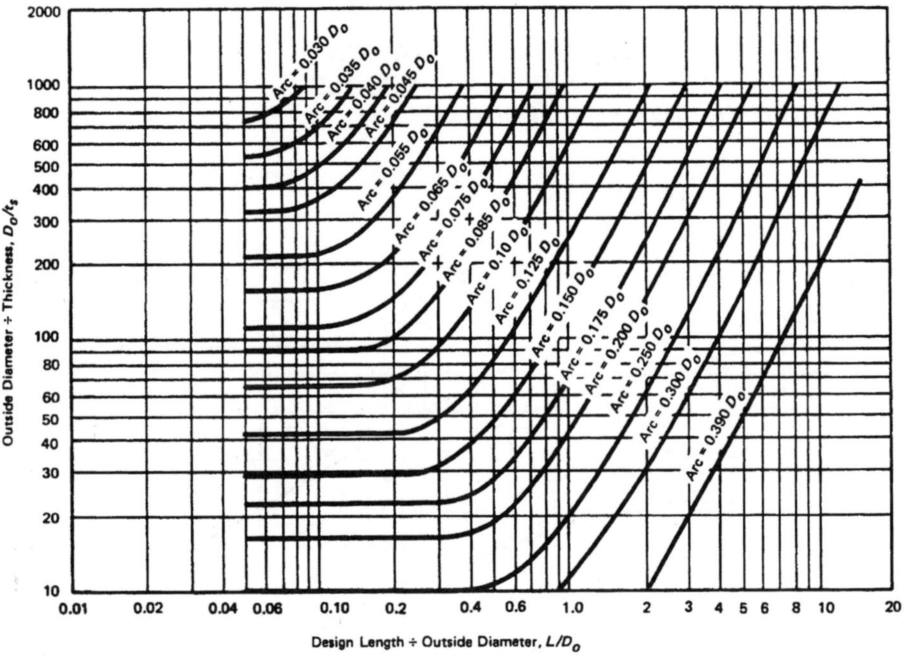

Figure 21.2 (ASME Figure UG 29.2) Maximum are of shell left unsupported because of gap in stiffening ring of cylindrical shell under external pressure.

5 UW 33: ALIGNMENT TOLERANCE

Refer to Chapter 3, Section 7.2 and Chapter 4, Section 8.1.

(a) Alignment of sections at the edges to be butt welded

Table 21.1 UW 33: Tolerance For Alignment of Shells

Section thickness (in.)	Joint Categories	
	A	B, C, and D
Up to 1/2, incl.	1/4t	1/4t
Over 1/2 to 3/4, incl.	1/8 in.	1/4t
Over 3/4 to 1 1/2, incl.	1/8 in.	3/16 in.
Over 1 1/2 to 2, incl.	1/8 in.	1/8t
Over 2	Lesser of 1/16t or 3/8 in.	Lesser of 1/8t or 3/4 in.

shall be such that the maximum offset is not greater than the applicable amount for the welded joint category (see UW 3) under considerations, as listed in Table UW 33. The section thickness t is the nominal thickness of the thinner section of the joint.

(b) Any offset within the allowable tolerance provided above shall be faired at a three-to-one taper over the width of the finished weld, or if necessary, by adding additional weld metal beyond what would otherwise be the edge of the weld. Such additional weld metal build up shall be subject to the requirements of UW 42.

6 UW 11: RADIOGRAPHIC AND ULTRASONIC EXAMINATION

Refer to Chapter 3, Section 8.3 and Chapter 4, Section 7.16.

(a) *Full radiography*. The following welded joints shall be examined radiographically for their full length in the manner prescribed in UW 51:

(1) All butt welds in the shell and heads of vessels used to contain lethal substances (see UW 2(a)).

(2) All butt welds in vessels in which the nominal thickness (see (g) below) at the welded joint exceeds 1 1/2 in. (38 mm), or exceeds the lesser thick-

nesses prescribed in UCS 57, UNF 57, UHA 33, UCL 35, or UCL 36 for the materials covered therein, or as otherwise prescribed in UHT 57, ULW 51, ULW 52(d), ULW 54, or ULT 57; however, except as required by UHT 57(a), categories B and C butt welds in nozzles and communicating chambers that neither exceed NPS 10 nor 1$^{1}/_{8}$ in. (29 mm) wall thickness do not require any radiographic examination.

(3) All butt welds in the shell and heads of unfired steam boilers having design pressures exceeding 50 psi (345 kPa) (see UW 2(c)).

(4) All butt welds in nozzles, communicating chambers, etc., attached to vessel sections or heads that are required to be fully radiographed under (1) or (3) above; however, except as required by UHT 57(a), categories B and C butt welds in nozzles and communicating chambers that neither exceed NPS 10 nor 1$^{1}/_{8}$ in. (29 mm) wall thickness do not require any radiographic examination.

(5) All category A and D butt welds in vessel sections and heads where the design of the joint or part is based on a joint efficiency permitted by UW 12(a), in which case:

(a) Category A and B welds connecting the vessel sections or heads shall be of Type No. (1) or Type No. (2) of Table UW 12.

(b) Category B or C butt welds [but not including those in nozzles or communicating chambers except as required in (2) above] which intersect the category A butt welds in vessel sections or heads or connect seamless vessel sections or heads shall, as a minimum, meet the requirements for spot radiography in accordance with UW 52. Spot radiographs required by this paragraph shall not be used to satisfy the spot radiography rules as applied to any other weld increment.

(6) All but welds joined by electrogas welding with any single pass greater than 1½ in. (38 mm) and all butt welds joined by electroslag welding.

(7) Ultrasonic examination in accordance with UW 53 may be substituted for radiography for the final closure seam of a pressure vessel if the construction of the vessel does not permit interpretable radiographs in accordance with Code requirements. The absence of suitable radiographic equipment shall not be justification for such substitution.

(8) Exemptions from radiographic examination for certain welds in nozzles and communicating chambers as described in (2), (4), and (5) above take precedence over the radiographic requirements of Subsection C of this Division.

(b) *Spot radiography.* Except as required in (a)(5)(b) above, butt welded joints made in accordance with Type No. (1) or (2) of Table UW 12 which are not required to be fully radiographed by (a) above, may be examined by spot radiography. Spot radiography shall be in accordance with UW 52. If spot radiography is specified for the entire vessel, radiographic examination is not required of category B and C butt welds in nozzles and communicating chambers that exceed neither NPS 10 nor 1⅛ in. (29 mm) wall thickness. *Note*: This requirement specifies spot radiography for butt welds of Type No. (1) or (2) that are used in a vessel, but does not preclude the use of fillet and/or corner welds permitted by other paragraphs, such as for nozzle and manhole attachments, welded stays, flat heads, etc., which need not be spot radiographed.

(c) *No radiography.* Except as required in (a) above, no radiographic examination of welded joints is required when the vessel or vessel part is designed for external pressure only, or when the joint design complies with UW 12(c)

(d) Electrogas welds in ferritic materials with any single pass greater than 1½ in. (38 mm) and electroslag welds in ferritic materials shall be ultrasonically examined throughout their entire length in accordance with the requirements of Appendix 12. This ultrasonic examination shall be done following the grain refining (austenitizing) heat treatment or post weld heat treatment.

(e) In addition to the requirements in (a) and (b) above, all welds made by the electron beam process shall be ultrasonically examined for their entire length in accordance with the requirements of Appendix 12.

(f) When radiography is required for a welded joint in accordance with (a) and (b) above, and the weld is made by the inertia and continuous drive friction welding processes, the welded joints shall also be ultrasonically examined for their entire length in accordance with Appendix 12.

(g) For radiographic and ultrasonic examination of butt welds, the definition of the nominal thickness at the welded joint under consideration shall be the nominal thickness of the thinner of the two parts joined. Nominal thickness is defined in 3-2.

7 UW 50: NONDESTRUCTIVE EXAMINATION OF WELDS ON PNEUMATICALLY TESTED VESSELS

Refer to Chapter 3, Section 8.3.

On welded pressure vessels to be pneumatically tested in accordance with UG 100, the full length of the following welds shall be examined for the purpose of detecting cracks:

(a) All welds around openings
(b) All attachment welds, including welds attaching nonpressure parts to pressure parts, having a throat thickness greater than 1/4 in. (6 mm)

8 UW 51; RADIOGRAPHIC AND RADIOSCOPIC EXAMINATION OF WELDED JOINTS

Refer to Chapter 3, Section 8.3 and Chapter 4, Section 7.16.

(a) All welded joints to be radiographed shall be examined in accordance with Article 2 of Section V except as specified below.

(1) A complete set of radiographs and records, as described in T 291 and T 292 of Article 2 of Section V, for each vessel or vessel part shall be retained by the manufacturer until the manufacturer's data report has been signed by the inspector.

(2) The manufacturer shall certify that personnel performing and evaluating radiographic examinations required by this Division have been qualified and certified in accordance with their employer's written practice. SNT-TC-1A shall be used as a guideline for employers to establish their written practice for qualification and certification of their personnel. Alternatively, the ASNT Central Certification Program (ACCP) may be used to fulfill this examination and demonstration requirements of SNT-TC-1A and the employer's written practice. Provisions for training, experience, qualification, and certification of NDE personnel shall be described in the manufacturer's quality control system [see Appendix 10].

(3) The requirements of T 285 of Article 2 of Section V are to be used only as a guide. Final acceptance of radiographs shall be based on the ability to see the prescribed penetrameter image and the specified hole or the designated wire of a wire penetrameter.

(b) Indications shown on the radiographs of welds and characterized as imperfections are unacceptable under the following conditions and shall be repaired as provided in UW 38, and the repair radiographed to UW 51 or, at the option of the manufacturer, ultrasonically examined in accordance with the method described in Appendix 12 and the standards specified in this paragraph, provided the defect has been confirmed by the ultrasonic examination to the satisfaction of the authorized inspector prior to making the repair. For material thicknesses in excess of 1 in., the concurrence of the user shall be obtained. This ultrasonic examination shall be noted under remarks on the manufacturer's data report form:

(1) Any indication characterized as a crack or zone of incomplete fusion or penetration.

(2) Any other elongated indication on the radiograph which has length greater than

(a) $1/4$ in. for t up to $3/4$ in.

(b) $1/3t$ for t from $3/4$ in. to $2^1/4$ in.

(c) $3/4$ in. for t over $2^1/4$ in., where t = the thickness of the weld excluding any allowable reinforcement. For a butt weld joining two members having different thicknesses at the weld, t is the thinner of these two thicknesses. If a full penetration weld includes a fillet weld, the thickness of the throat of the fillet shall be included in t.

(3) Any group of aligned indications that have an aggregate length greater than t in a length of $12t$, except when the distance between the successive imperfections exceeds $6L$, where L is the length of the longest imperfection in the group.

(4) Rounded indications in excess of that specified by the acceptance standards given in Appendix 4.

(c) All welded joints to be examined by real time radioscopic examination shall be examined in accordance with Appendix II of Article 2 of Section V as specified below.

(1) A complete set of records, as described in II-292, shall be evaluated by the manufacturer prior to being presented to the inspector. Imperfections listed in UW 51(b)(1), (2), (3), and (4) are unacceptable and shall be repaired as provided in UW 38 and the repair re-examined by either film or real time radioscopic examination. Records shall be retained by the manufacturer until the data report has been signed by the inspector.

(2) Provisions for training, experience, qualification, and certification of personnel responsible for equipment set-up, calibration, operation, and eval-

uation of examination data shall be described in the manufacturer's quality control system [see Appendix 10].

(3) The use of real time radioscopic examination shall be noted under remarks on the Manufacturer's Data Report.

9 UW 52; SPOT EXAMINATION OF WELDED JOINTS

Refer to Chapter 3, Section 8.3 and Chapter 4, Section 7.16.

Note: Spot radiographing of a welded joint is recognized as an effective inspection tool. The spot radiography rules are also considered to be an aid to quality control. Spot radiographs made directly after a welder or an operator has completed a unit of weld proves that the work is or is not being done in accordance with a satisfactory procedure. If the work is unsatisfactory, corrective steps can then be taken to improve the welding in the subsequent units, which unquestionably will improve the weld quality.

Spot radiography in accordance with these rules will not ensure a fabrication product of a predetermined quality level throughout. It must be realized that an accepted vessel under these spot radiography rules may still contain defects which might be disclosed on further examination. If all radiographically disclosed weld defects must be eliminated from a vessel, then 100% radiography must be employed.

(a) Butt welded joints which are to be spot radiographed shall be examined locally as provided herein.

(b) *Minimum extent of spot radiographic examination*:

(1) One spot shall be examined on each vessel for each 50 ft (15.2 m) increment of weld or fraction thereof for which a joint efficiency from column (b) of Table UW 12 is selected. However, for identical vessels, each with less than 50 ft (15.2 m) of weld for which a joint efficiency from column (b) of Table UW 12

is selected, 50 ft (15.2 m) increments of weld may be represented by one spot examination.

(2) For each increment of weld to be examined, a sufficient number of spot radiographs shall be taken to examine the welding of each welder or welding operator. Under conditions where two or more welders or welding operators make weld layers in a joint, or on the two sides of a double-welded butt joint, one spot may represent the work of all welders or welding operators.

(3) Each spot examination shall be made as soon as practicable after completion of the increment of weld to be examined. The location of the spot shall be chosen by the inspector after completion of the increment of welding to be examined, except that when the inspector has been notified in advance and cannot be present or otherwise make the selection, the fabricator may exercise his own judgment in selecting the spots.

(4) Radiographs required at specific locations to satisfy the rules of other paragraphs, such as UW 9(d), UW 11(a)(5)(b), and UW 14(b), shall not be used to satisfy the requirements for spot radiography.

(c) *Standards for spot radiographic examination.* Spot examination by radiography shall be made in accordance with the technique prescribed in UW 51(a). The minimum length of the spot radiograph shall be 6 in. Spot radiographs may be retained or discarded by the manufacturer after acceptance of the vessel by the inspector. The acceptability of welds examined by spot radiography shall be judged by the following standards:

(1) Welds in which indications are characterized as cracks or zones of incomplete fusion or penetration shall be unacceptable.

(2) Welds in which indications are characterized as slag inclusions or cavities shall be unacceptable if the length of any such indication is greater than

$^2/_3t$, where t is the thickness of the weld excluding any allowable reinforcement. For a butt weld joining two members having different thicknesses at the weld, t is the thinner of these two thicknesses. If a full penetration weld includes a fillet weld, the thickness of the throat of the fillet shall be included in t. If several indications within the above limitations exist in line, the welds shall be judged acceptable if the sum of the longest dimensions of all such indications is not more than t in a length of $6t$ (or proportionately for radiographs shorter than $6t$) and if the longest indications considered are separated by at least $3L$ of acceptable weld metal, where L is the length of the longest indication. The maximum length of acceptable indications shall be $^3/_4$ in. (19 mm). Any such indications shorter than $^1/_4$ in. (6 mm) shall be acceptable for any plate thickness.

(3) Rounded indications are not a factor in the acceptability of welds not required to be fully radiographed.

(d) *Evaluation and retests*:

(1) When a spot, radiographed as required in (b)(1) or (b)(2) above, is acceptable in accordance with (c)(1) and (c)(2) above, the entire weld increment represented by this radiograph is acceptable.

(2) When a spot, radiographed as required in (b)(1) or (b)(2) above, has been examined and the radiograph discloses welding which does not comply with the minimum quality requirements of (c)(1) or (c)(2) above, two additional spots shall be radiographically examined in the same weld increment at locations away from the original spot. The locations of these additional spots shall be determined by the inspector or fabricator as provided for the original spot examination in (b)(3) above.

(a) If the two additional spots examined show welding which meets the minimum quality requirements of (c)(1) and (c)(2) above, the entire weld increment represented by the three radiographs is acceptable provided the defects disclosed by the first of the three radiographs are removed and the area repaired by welding. The weld-repaired area shall be radiographically examined in accordance with the foregoing requirements of UW 52.

(b) If either of the two additional spots examined shows welding which does not comply with the minimum quality requirements of (c)(1) or (c)(2) above, the entire increment of weld represented shall be rejected. The entire rejected weld shall be removed and the joint shall be rewelded or, at the fabricator's option, the entire increment of weld represented shall be completely radiographed and only defects need be corrected.

(c) Repair welding shall be performed using a qualified procedure and in a manner acceptable to the inspector. The rewelded joint, or the weld-repaired areas, shall be spot radiographically examined at one location in accordance with the foregoing requirements of UW 52.

10 ASME APPENDIX 4: ROUNDED INDICATIONS CHARTS, ACCEPTANCE STANDARD FOR RADIOGRAPHICALLY DETERMINED ROUNDED INDICATIONS IN WELDS

Refer to Chapter 3, Section 8.3 and Chapter 4, Section 7.16.

4-1 Applicability of These Standards

These standards are applicable to ferritic, austenitic, and nonferrous materials.

4-2 Terminology

(a) *Rounded indications.* Indications with a maximum length of three times the width or less on the radiograph are defined as rounded indications. These indications may be circular, elliptical, conical, or irregular in shape and may have tails. When evaluating the size of an indication, the tail shall be included. The indication may be from any imperfection in the weld, such as porosity, slag, or tungsten.

(b) *Aligned indications.* A sequence of four or more rounded indications shall be considered to be aligned when they touch a line parallel to the length of the weld drawn through the center of the two outer rounded indications.

(c) *Thickness*(t). t is the thickness of the weld, excluding any allowable reinforcement. For a butt weld joining two members having different thicknesses at the weld, t is the thinner of these two thicknesses. If a full penetration weld includes a fillet weld, the thickness of the throat of the fillet shall be included in t.

4-3 Acceptance Criteria

(a) *Image density.* Density within the image of the indication may vary and is not a criterion for acceptance or rejection.

(b) *Relevant indications.* (See Table A4-1 for examples.) Only those rounded indications which exceed the following dimensions shall be considered relevant:

$^1/_{10}\ t$ for t less than $^1/_8$ in. (3.2 mm).
$^1/_{64}$ in. for t from $^1/_8$ to $^1/_4$ in. (3.2 to 6 mm), incl.
$^1/_{32}$ in. for t greater than $^1/_4$ to 2 in. (6 to 51 mm), incl.
$^1/_{16}$ in. for t greater than 2 in. (51 mm)

(c) *Maximum size of rounded indication.* (See Table A4-1 for examples.) The maximum permissible size of any indication shall be $^1/_4\ t$, or $^5/_{32}$ in. (4 mm), whichever is smaller; except that an isolated indication separated from an adjacent indication by 1 in. (25 mm) or more may be $^1/_3\ t$ or $^1/_4$ in. (6 mm), whichever is less. For t greater than 2 in. (51 mm), the maximum permissible size of an isolated indication shall be increased to $^3/_8$ in. (10 mm).

(d) *Aligned rounded indications.* Aligned rounded indications are acceptable when the summation of the diameters of the indications is less than t in a length of 12 t. See Figure

A4-1. The length of groups of aligned rounded indications and the spacing between the groups shall meet the requirements of Figure A4-2.

(e) *Spacing.* The distance between adjacent rounded indications is not a factor in determining acceptance or rejection, except as required for isolated indications or groups of aligned indications.

(f) *Rounded indication charts.* The rounded indications characterized as imperfections shall not exceed those shown in the charts. The charts in Figures A4-3 through A4-8 illustrate various types of assorted, randomly dispersed and clustered rounded indications for different weld thicknesses greater than 1/8 in. (3.2 mm). These charts represent the maximum acceptable concentration limits for rounded indications. The charts for each thickness range represent full-scale 6-in. radiographs and shall not be enlarged or reduced. The distributions shown are not necessarily the patterns that may appear on the radiograph, but are typical of the concentration and size of indications permitted.

(g) *Weld thickness* t *less than 1/8 in. (3.2 mm).* For *t* less than 1/8 in. (3.2 mm), the maximum number of rounded indications shall not exceed 12 in a 6-in. (152-mm) length of weld. A proportionally fewer number of indications shall be permitted in welds less than 6 in. (152 mm) in length.

(h) *Clustered indications.* The illustrations for clustered indications show up to four times as many indications in a local area, as that shown in the illustrations for random indications. The length of an acceptable cluster shall not exceed the lesser of 1 in. (25 mm) or 2*t*. Where more than one cluster is present, the sum of the lengths of the clusters shall not exceed 1 in. (25 mm) in a 6-in. (152-mm) length weld.

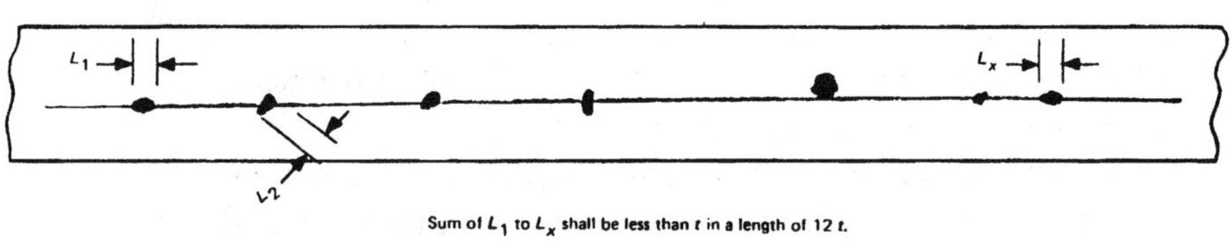

Figure A4-1 Aligned rounded indications.

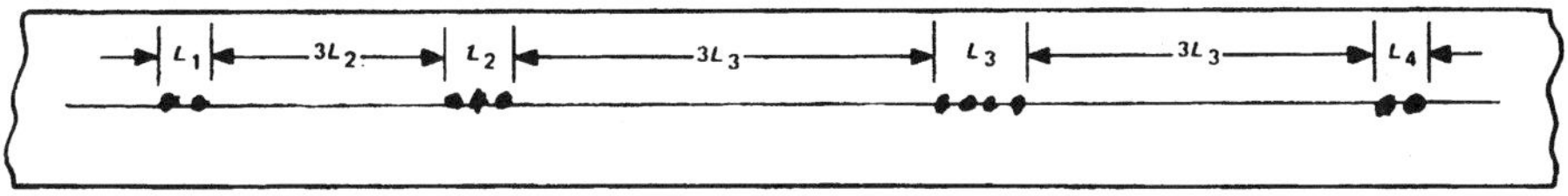

The sum of the group lengths shall be less than t in a length of $12t$.

Maximum Group Length	Minimum Group Spacing
L = 1/4 in. for t less than 3/4 in.	$3L$ where L is the length of the longest adjacent group being evaluated.
L = 1/3 t for t 3/4 in. to 2-1/4 in.	
L = 3/4 in. for t greater than 2-1/4 in.	

Figure A4-2 Groups of aligned rounded indications.

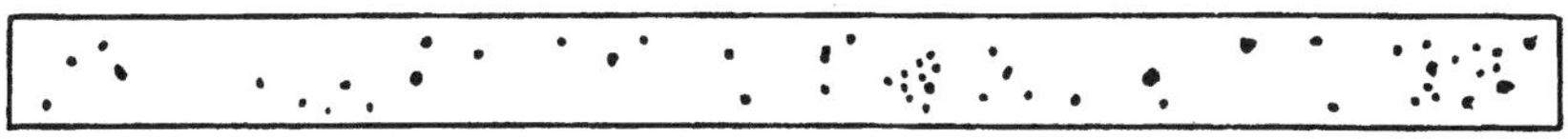

RANDOM ROUNDED INDICATIONS

Typical concentration and size permitted in any 6 in. length of weld

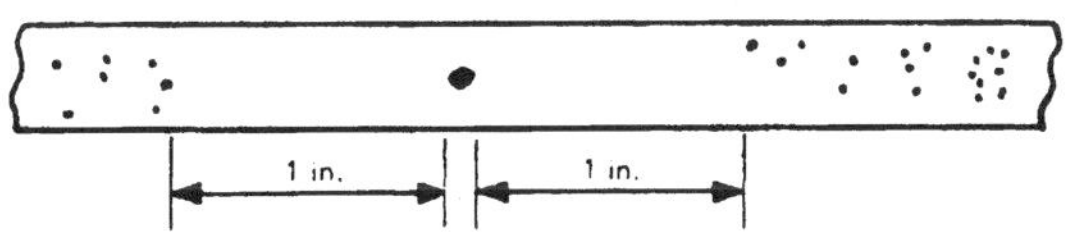

ISOLATED INDICATION
Maximum size per Table 4-1

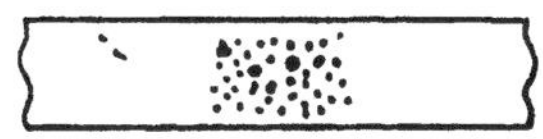

CLUSTER

Figure A4-3 Charts for t equal to $^1/_8$ in. to $^1/_4$ in., inclusive.

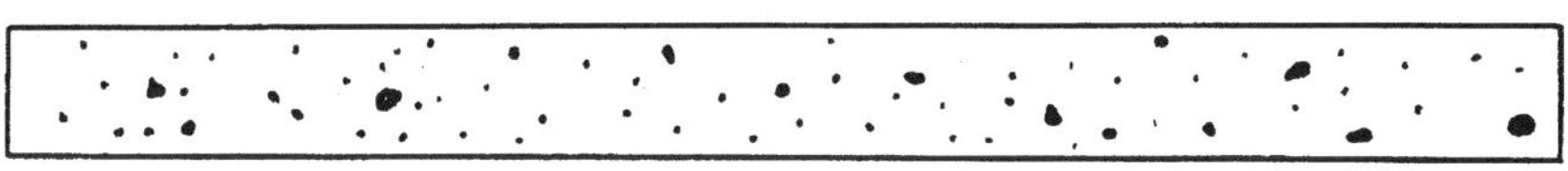

RANDOM ROUNDED INDICATIONS

Typical concentration and size permitted in any 6 in. length of weld

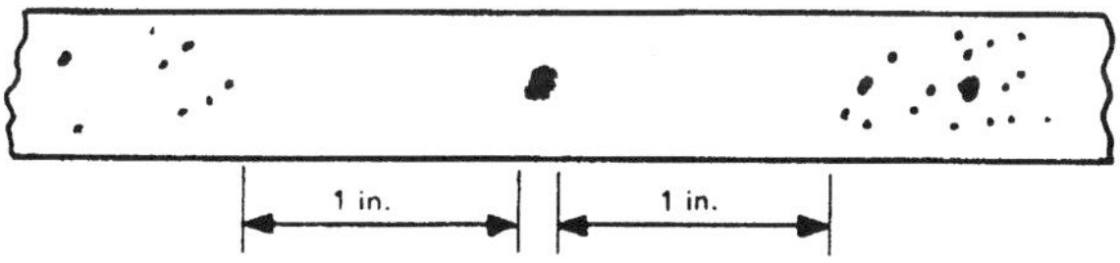

ISOLATED INDICATION
Maximum size per Table 4-1

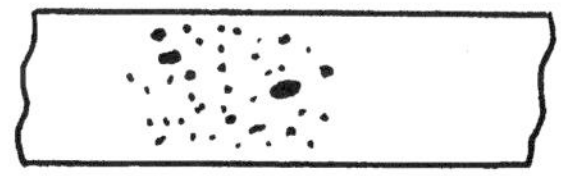

CLUSTER

Figure A4-4 Charts for t over $^1/_4$ in. to $^3/_8$ in., inclusive.

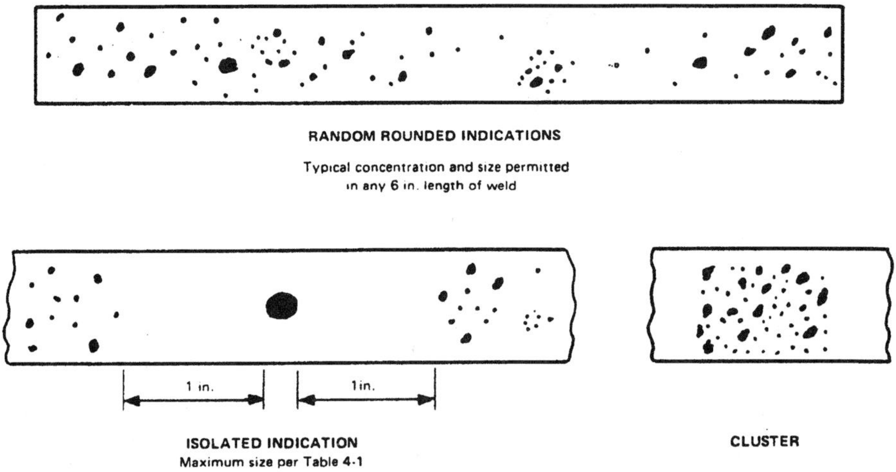

Figure A4-5 Charts for t over 3/8 in. to 3/4 in., inclusive.

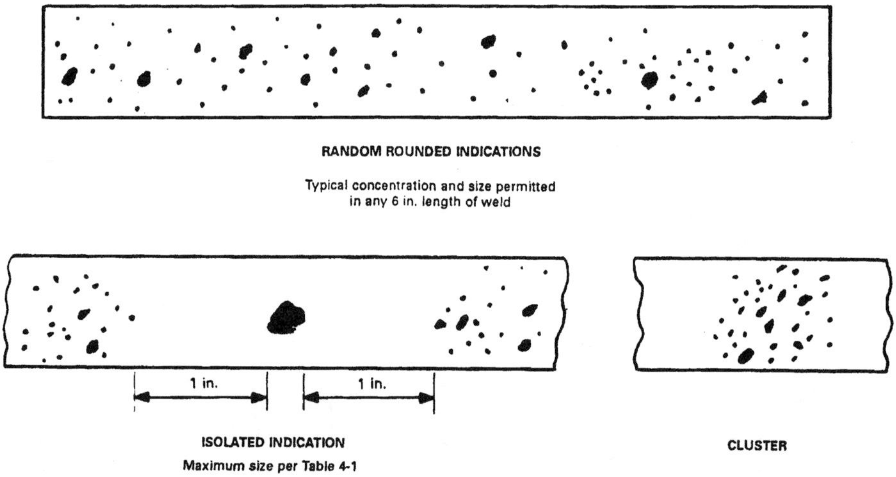

Figure A4-6 Charts for t over 3/4 in. to 2 in., inclusive.

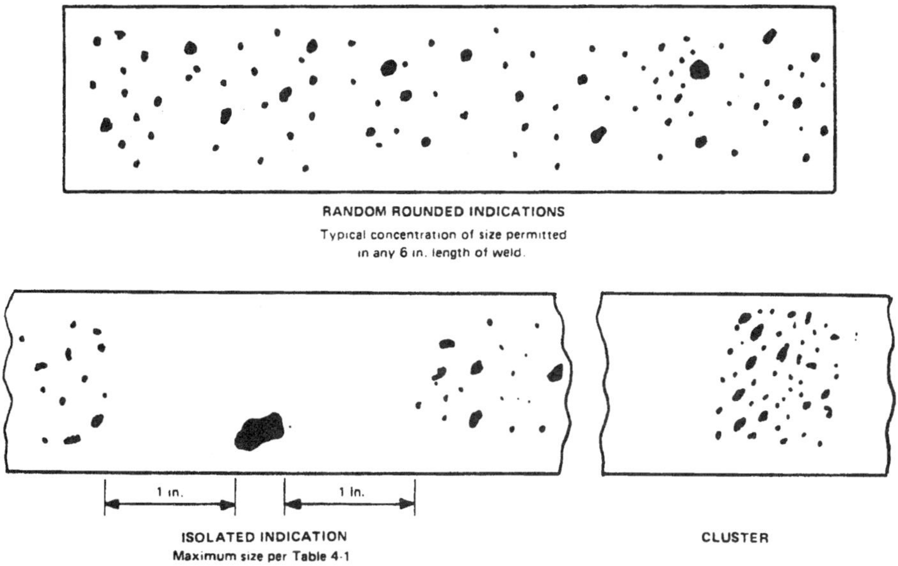

Figure A4-7 Charts for t over 2 in. to 4 in., inclusive.

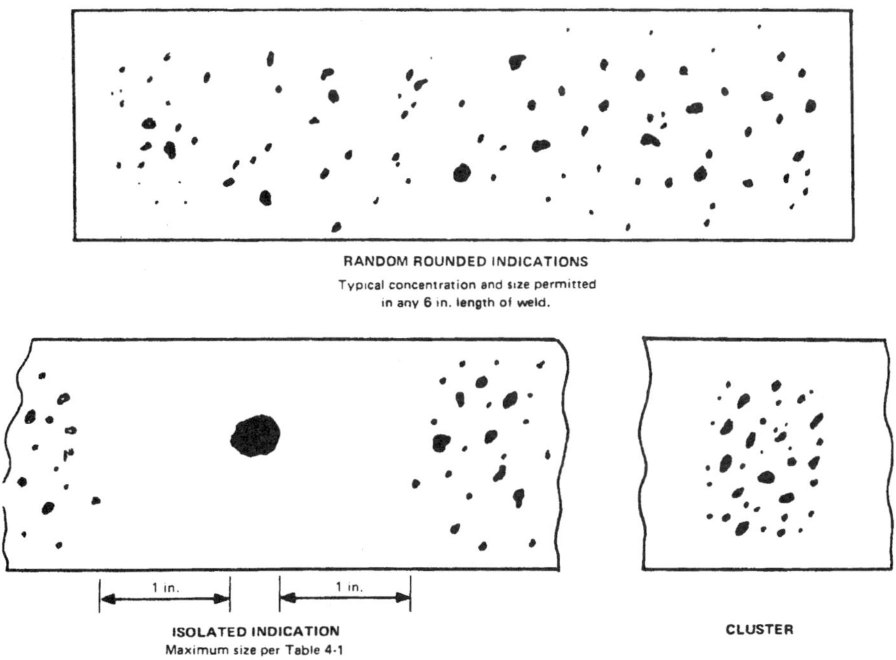

Figure A4-8 Charts for t over 4 in.

Table A4-1

Thickness t (in.)	Maximum size of acceptable rounded indication (in.) Random	Isolated	Maximum size of nonrelevant indication (in.)
Less than 1/8	$\frac{1}{4}t$	$\frac{1}{3}t$	$\frac{1}{10}t$
1/8	0.031	0.042	0.015
3/16	0.047	0.063	0.015
1/4	0.063	0.083	0.015
5/16	0.078	0.104	0.031
3/8	0.091	0.125	0.031
7/16	0.109	0.146	0.031
1/2	0.125	0.168	0.031
9/16	0.142	0.188	0.031
5/8	0.156	0.210	0.031
11/16	0.156	0.230	0.031
3/4 to 2, incl.	0.156	0.250	0.031
Over 2	0.156	0.375	0.063

Notes: (1) This table contains examples only.

11 UW 53: TECHNIQUE FOR ULTRASONIC EXAMINATION OF WELDED JOINTS

Refer to Chapter 3, Section 8.3 and Chapter 4, Section 7.16.

Ultrasonic examination of welded joints when required or permitted by other paragraphs of this Division shall be performed in accordance with Appendix 12 and shall be evaluated to the acceptance standards specified in Appendix 12. The written examination procedure shall be available to the inspector and shall be proven by actual demonstration to the satisfaction of the inspector to be capable of detecting and locating imperfections described in this Division.

12 ASME APPENDIX 12: ULTRASONIC EXAMINATION OF WELDS (UT)

Refer to Chapter 3, Section 8.3 and Chapter 4, Section 7.16.

12-1 Scope

(a) This Appendix describes methods which shall be employed when ultrasonic examination of welds is specified in this Division.

(b) Article 5 of Section V shall be applied for detail requirements in methods, procedures, and qualifications unless otherwise specified in this Appendix.

(c) Ultrasonic examination shall be performed in accordance with a written procedure, certified by the manufacturer to be in accordance with the requirements of T 150 of Section V.

12-2 Certification of Competence of Nondestructive Examiner

The manufacturer shall certify that personnel performing and evaluating ultrasonic examinations required by this Division have been qualified and certified in accordance with their employer's written practice. SNT-TC-1A shall be used as a guideline for employers to establish their written practice for qualification and certification of their personnel. Provisions for training, experience, qualification, and certification of NDE personnel shall be described in the manufacturer's quality control system [see Appendix 10].

12-3 Acceptance/Rejection Standards

These standards shall apply unless other standards are specified for specific applications within this Division.

Imperfections which produce a response greater than 20% of the reference level shall be investigated to the extent that the operator can determine the shape, identity, and location of all such imperfections and evaluate them in terms of the acceptance standards given in (a) and (b) below.

(a) Indications characterized as cracks, lack of fusion, or incomplete penetration are unacceptable regardless of length.
(b) Other imperfections are unacceptable if the indications exceed the reference level amplitude and have lengths which exceed
  (1) 1/4 in. (6 mm) for t up to 3/4 in. (19 mm)
  (2) 1/3t for t from 3/4 to 2 1/4 in. (19 to 57 mm)
  (3) 3/4 in. for t over 2 1/4 in. (57 mm)

  where t is the thickness of the weld excluding any allowable reinforcement. For a butt weld joining two members having different thicknesses at the weld, t is the thinner of these two thicknesses. If a full penetration weld includes a fillet weld, the thickness of the throat of the fillet shall be included in t.

12-4 Report of Examination

The manufacturer shall prepare a report of the ultrasonic examination and a copy of this report shall be retained by the manufacturer until the manufacturer's data report has been signed by the inspector. The report shall contain the information required by Section V. In addition, a record of repaired areas shall be noted as well as the results of the re-examination of the repaired areas. The manufacturer shall also maintain a record of all reflections from uncorrected areas having responses that exceed 50% of the reference level. This record shall locate each area, the response level, the dimensions, the depth below the surface, and the classification.

13 ASME APPENDIX 6: METHODS FOR MAGNETIC PARTICLE EXAMINATION (MT)

Refer to Chapter 3, Section 8.3 and Chapter 4, Section 7.16.

6-1 Scope

(a) This Appendix provides for procedures which shall be followed whenever magnetic particle examination is specified in this Division.
(b) Article 7 of Section V shall be applied for the detail requirements in methods and procedures, and the additional requirements specified within this Appendix.
(c) Magnetic particle examination shall be performed in accordance with a written procedure, certified by the manufacturer to be in accordance with the requirements of T 150 of Section V.

6-2 Certification of Competency for Nondestructive Examination Personnel

The manufacturer shall certify that each magnetic particle examiner meets the following requirements:

(a) He has vision, with correction if necessary, to enable him to read a Jaeger Type No. 2 Standard Chart at a distance of not less than 12 in. and is capable of distinguishing and differentiating contrast between colors used. These requirements shall be checked annually.
(b) He is competent in techniques of the magnetic particle examination method for which he is certified, including making the examination, and interpreting and evaluating the results, except that where the examination method consists of more than one operation, he may be certified as being qualified only for one or more of these operations.

6-3 Evaluation of Indications

Indications will be revealed by retention of magnetic particles. All such indications are not necessarily imperfections, however, since excessive surface roughness, magnetic permeability variations (such as at the edge of heat affected zones), etc., may produce similar indications. An indication is the evidence of a mechanical imperfection. Only indications which have any dimension greater than 1/16 in. (1.6 mm) shall be considered relevant.

(a) A linear indication is one having a length greater than three times the width.
(b) A rounded indication is one of circular or elliptical shape with a length equal to or less than three times its width.
(c) Any questionable or doubtful indications shall be reexamined to determine whether or not they are relevant.

6-4 Acceptance Standards

These acceptance standards shall apply unless other more restrictive standards are specified for specific materials or applications within this Division.

All surfaces to be examined shall be free of

(a) Relevant linear indications
(b) Relevant rounded indications greater than 3/16 in. (4.8 mm)
(c) Four or more relevant rounded indications in a line separated by 1/16 in. (1.6 mm) or less, edge to edge
(d) An indication of an imperfection may be larger than the imperfection that causes it; however, the size of the indication is the basis for acceptance evaluation.

6-5 Repair Requirements

The defect shall be removed or reduced to an imperfection of acceptable size. Whenever an imperfection is removed by chipping or grinding and subsequent repair by welding is not required, the excavated area shall be blended into the surrounding surface so as to avoid sharp notches, crevices, or corners. Where welding is required after removal of an imperfection, the area shall be cleaned and welding performed in accordance with a qualified welding procedure.

(a) *Treatment of indications believed nonrelevant.* Any indication which is believed to be nonrelevant shall be regarded as an imperfection unless it is shown by re-examination by the same method or by the use of other nondestructive methods and/or by surface conditioning that no unacceptable imperfection is present.

(b) *Examination of areas from which imperfections have been removed.* After a defect is thought to have been removed and prior to making weld repairs, the area shall be examined by suitable methods to ensure it has been removed or reduced to an acceptably sized imperfection.

(c) *Re-examination of repair areas.* After repairs have been made, the repaired area shall be blended into the surrounding surface so as to avoid sharp notches, crevices, or corners and re-examined by the magnetic particle method and by all other methods of examination that were originally required for the affected area, except that, when the depth of repair is less than the radiographic sensitivity required, reradiography may be omitted.

14 ASME APPENDIX 8: METHODS FOR LIQUID PENETRANT EXAMINATION (PT)

Refer to Chapter 3, Section 8.3 and Chapter 4, Section 7.16. Note: Satisfactory application of this method of examination requires special skills in the techniques involved and in interpresting the results. The requirements specified herein presume application by suitably experienced personnel.

8-1 Scope

(a) This Appendix describes methods which shall be employed whenever liquid penetrant examination is specified in this Division.

(b) Article 6 of Section V shall be applied for detail requirements in methods, procedures, and qualifications unless specified within this Appendix.

(c) Liquid penetrant examination shall be performed in accordance with a written procedure, certified by the manufacturer to be in accordance with the requirements of T 150 of Section V.

8-2 Certification of Competency of Nondestructive Examination Personnel

The manufacturer shall certify that each liquid penetrant examiner meets the following requirements:

(a) He has vision, with correction if necessary, to enable him to read a Jaeger Type No. 2 Standard Chart at a distance of not less than 12 in. (305 mm) and is capable of distinguishing and differentiating contrast between colors used. These requirements shall be checked annually.

(b) He is competent in techniques of the liquid penetrant examination method for which he is certified, including making the examination and interpreting and evaluating the results, except that, where the examination method consists of more than one operation, he may be certified as being qualified only for one or more of these operations.

8-3 Evaluation of Indications

An indication is the evidence of a mechanical imperfection. Only indications with major dimensions greater than $^1/_{16}$ in. shall be considered relevant.

(a) A linear indication is one having a length greater than three times the width.

(b) A rounded indication is one of circular or elliptical shape with the length equal to or less than three times the width.

(c) Any questionable or doubtful indications shall be reexamined to determine whether or not they are relevant.

8-4 Acceptance Standards

These acceptance standards shall apply unless other more restrictive standards are specified for specific materials or applications within this Division.

All surfaces to be examined shall be free of

(a) Relevant linear indications

(b) Relevant rounded indications greater than 3/16 in. (4.8 mm)

(c) Four or more relevant rounded indications in a line separated by 1/16 in. (1.6 mm) or less, edge to edge

(d) An indication of an imperfection may be larger than the imperfection that causes it; however, the size of the indication is the basis for acceptance evaluation.

8-5 Repair Requirements

Unacceptable imperfections shall be repaired and re-examination made to assure removal or reduction to an acceptable size. Whenever an imperfection is repaired by chipping or grinding and subsequent repair by welding is not required, the excavated area shall be blended into the surrounding surface so as to avoid sharp notches, crevices, or corners. Where welding is required after repair of an imperfection, the area shall be cleaned and welding performed in accordance with a qualified welding procedure.

(a) *Treatment of indications believed nonrelevant.* Any indication which is believed to be nonrelevant shall be regarded as an imperfection unless it is shown by reexamination by the same method or by the use of other nondestructive methods and/or by surface conditioning that no unacceptable imperfection is present.

(b) *Examination of areas from which defects have been removed.* After a defect is thought to have been removed and prior to making weld repairs, the area shall be examined by suitable methods to ensure it has been removed or reduced to an acceptably sized imperfection.

(c) *Re-examination of repair areas.* After repairs have been made, the repaired area shall be blended into the surrounding surface so as to avoid sharp notches, crevices, or corners and re-examined by the liquid penetrant method and by all other methods of examination that were originally required for the affected area, except that, when the depth of repair is less than the radiographic sensitivity required, reradiography may be omitted.

15 ASME APPENDIX R: PREHEATING

Refer to Chapter 4, section 7.7.

Preheating may be employed during welding to assist in completion of the welded joint. The need for and temperature of preheat are dependent on a number of factors, such as the chemical analysis, degree of restraint of the parts being joined, elevated physical properties, and heavy thicknesses. Mandatory rules for preheating are, therefore, not given in this Division except as required in the footnotes that provide for exemptions to postweld heat treatment in Tables UCS 56 and UHA 32. Some practices used for preheating are given below as a general guide for the materials listed by P. numbers in Section IX. It is cautioned that the preheating temperatures listed below do not necessarily insure satisfactory completion of the welded joint and requirements for individual materials within the P. number listing may have preheating more or less restrictive than this general guide. The procedure specification for the material being welded specifies the minimum preheating requirements under Section IX weld procedure qualification requirements.

The heat of welding may assist in maintaining preheat temperatures after the start of welding and for inspection purposes, temperature checks can be made near the weld. The method or extent of application of preheat is not, therefore, specifically given. Normally when materials of two different P. number groups are joined by welding, the preheat used will be that of the material with the higher preheat specified on the procedure specification.

R-1 P. No. 1 Groups 1, 2, and 3

(a) 175°F (79°C) for material which has both a specified maximum carbon content in excess of 0.30% and a thickness at the joint in excess of 1 in. (25 mm)
(b) 50°F (10°C) for all other materials is this P. number.

R-2 P. No. 3 Group Nos. 1, 2, and 3

(a) 175°F (79°C) for material which has either a specified minimum tensile strength in excess of 70,000 psi (480 MPa) or a thickness at the joint in excess of 5/8 in. (16 mm)
(b) 50°F (10°C) for all other materials is this P. number.

R-3 P. No. 4 Group Nos. 1 and 2

(a) 250°F (121°C) for material which has either a specified minimum tensile strength in excess of 60,000 psi (410 MPa) or a thickness at the joint in excess of 1/2 in. (13 mm)
(b) 50°F (10°C) for all other materials is this P. number.

R-4 P. Nos. 5A and 5B Group No. 1

(a) 400°F (204°C) for material which has either a specified minimum tensile strength in excess of 60,000 psi

(410 MPa) or has both a specified minimum chromium content above 6.0% and a thickness at the joint in excess of ½ in. (13 mm)

(b) 300°F (149°C) for all other materials is these P. numbers.

R-5 P. No. 6 Group Nos. 1, 2, and 3

400°F (204°C)

R-6 P. No. 7 Group Nos. 1 and 2

None.

R-7 P. No. 8 Group Nos. 1 and 2

None.

R-8 P. No. 9 Groups

250°F (121°C) for P. no. 9A group no. 1 materials
300°F (149°C) for P. no. 9B group no. 1 materials

R-9 P. No. 10 Group

175°F (79°C) for P. no. 10A group no. 1 materials.
250°F (121°C) for P. no. 10B group no. 2 materials.
175°F (79°C) for P. no. 10C group no. 3 materials.
250°F (121°C) for P. no. 10F group no. 6 materials.
For P. no. 10C group no. 3 materials, preheat is neither required nor prohibited, and consideration shall be given to the limitation of interpass temperature for various thicknesses to avoid detrimental effects on the mechanical properties of heat treated material.
For P. no. 10D group no. 4 and P. no. 10E group no. 5 materials, 300°F (149°C) with interpass temperature maintained between 350 and 450°F (177 and 232°C).

R-10 P. No. 11 Group

(a) P. no. 11A Group

Group no. 1: none (see *Note*)
Group no. 2: same as for P. no. 5 (see *Note*)
Group no. 3: same as for P. no. 5 (see *Note*)
Group no. 4: 250°F (121°C)

(b) P. no. 11B Group

Group no. 1: same as for P. no. 3 (see *Note*)
Group no. 2: same as for P. no. 3 (see *Note*)
Group no. 3: same as for P. no. 3 (see *Note*)
Group no. 4: same as for P. no. 3 (see *Note*)
Group no. 5: same as for P. no. 3 (see *Note*)
Group no. 6: same as for P. no. 5 (see *Note*)
Group no. 7: same as for P. no. 5 (see *Note*)

Note: Consideration shall be given to the limitation of interpass temperature for various thicknesses to avoid detrimental effects on the mechanical properties of heat treated materials.

16 UW 35; FINISHED LONGITUDINAL AND CIRCUMFERENTIAL JOINTS

Refer to Chapter 4, Section 7.15.

(a) Butt welded joints shall have complete penetration and full fusion. As-welded surfaces are permitted; however, the surface of welds shall be sufficiently free from coarse ripples, grooves, overlaps, and abrupt ridges and valleys to permit proper interpretation of radiographic and other required nondestructive examinations. If there is a question regarding the surface condition of the weld when interpreting a radiographic film, the film shall be compared to the actual weld surface for determination of acceptability.

(b) A reduction in thickness due to the welding process is acceptable provided all of the following conditions are met:

(1) The reduction in thickness shall not reduce the material of the adjoining surfaces below the minimum required thickness at any point.

(2) The reduction in thickness shall not exceed 1/32 in. (0.8 mm) or 10% of the nominal thickness of the adjoining surface, whichever is less.

(c) When a single-welded butt joint is made by using a backing strip which is left in place [Type No. (2) of Table UW 12], the requirement for reinforcement applies only to the side opposite the backing strip.

(d) To assure that the weld grooves are completely filled so that the surface of the weld metal at any point does not fall below the surface of the adjoining base materials, weld metal may be added as reinforcement on each face of the weld. The thickness of the weld reinforcement on each face shall not exceed the following:

Material nominal thickness (in.)	Maximum reinforcement (in.)	
	Category B and C butt welds	Other welds
Less than 3/32	3/32	1/32
3/32 to 3/16, incl.	1/8	1/16
Over 3/16 to 1/2, incl.	5/32	3/32
Over 1/2 to 1, incl.	3/16	3/32
Over 1 to 2, incl.	1/4	1/8
Over 2 to 3, incl.	1/4	5/32
Over 3 to 4, incl.	1/4	7/32
Over 4 to 5, incl.	1/4	1/4
Over 5	5/16	5/16

17 UW 40; PROCEDURES FOR POSTWELD HEAT TREATMENT

Refer to Chapter 7, Section 1.

(a) The operation of postweld heat treatment shall be

performed in accordance with the requirements given in the applicable Part in Subsection C using one of the following procedures. In the procedures that follow, the soak band is defined as the volume of metal required to meet or exceed the minimum PWHT temperatures listed in Table UCS 56. As a minimum, the soak band shall contain the weld, heat affected zone, and a portion of base metal adjacent to the weld being heat treated. The minimum width of this volume is the widest width weld plus 1*t* or 2 in., whichever is less, on each side or end of the weld. The term *t* is the nominal thickness as defined in (f) below.

(1) Heating the vessel as a whole in an enclosed furnace: This procedure is preferable and should be used whenever practicable.

(2) Heating the vessel in more than one heat in a furnace, provided the overlap of the heated sections of the vessel is at least 5 ft (1.5 m): When this procedure is used, the portion outside of the furnace shall be shielded so that the temperature gradient is not harmful. The cross-section where the vessel projects from the furnace shall not intersect a nozzle or other structural discontinuity.

(3) Heating of shell sections and/or portions of vessels to postweld heat treat longitudinal joints or complicated welded details before joining to make the completed vessel: When the vessel is required to be postweld heat treated, and it is not practicable to postweld heat treat the completed vessel as a whole or in two or more heats as provided in (2) above, any circumferential joints not previously postweld heat treated may be thereafter locally postweld heat treated by heating such joints by any appropriate means that will assure the required uniformity. For such local heating, the soak band shall extend around the full circumference. The portion outside the soak band shall be protected so that the temperature gradient is not

harmful. This procedure may also be used to postweld heat treat portions of new vessels after repairs.

(4) Heating the vessel internally by any appropriate means and with adequate indicating and recording temperature devices to aid in the control and maintenance of a uniform distribution of temperature in the vessel wall. Previous to this operation, the vessel should be fully enclosed with insulating material, or the permanent insulation may be installed provided it is suitable for the required temperature. In this procedure the internal pressure should be kept as low as practicable, but shall not exceed 50% of the maximum allowable working pressure at the highest metal temperature expected during the postweld heat treatment period.

(5) Heating a circumferential band containing nozzles or other welded attachments that require postweld heat treatment in such a manner that the entire band shall be brought up uniformly to the required temperature and held for the specified time. Except as modified in this paragraph below, the soak band shall extend around the entire vessel, and shall include the nozzle or welded attachment. The circumferential soak band width may be varied away from the nozzle or attachment weld requiring PWHT, provided the required soak band around the nozzle or attachment weld is heated to the required temperature and held for the required time. As an alternative to varying the soak band width, the temperature within the circumferential band away from the nozzle or attachment may be varied and need not reach the required temperature, provided the required soak band around the nozzle or attachment weld is heated to the required temperature, held for the required time, and the temperature gradient is not harmful

throughout the heating and cooling cycle. The portion of the vessel outside of the circumferential soak band shall be protected so that the temperature gradient is not harmful. This procedure may also be used to postweld heat treat portions of vessels after repairs.

(6) Heating the circumferential joints of pipe or tubing by any appropriate means using a soak band that extends around the entire circumference. The portion outside the soak band shall be protected so that the temperature gradient is not harmful. The proximity to the shell increases thermal restraint, and the designer should provide adequate length to permit heat treatment without harmful gradients at the nozzle attachment or heat a full circumferential band around the shell, including the nozzle.

(7) Heating a local area around nozzles or welded attachments in the larger radius sections of a double-curvature head or a spherical shell or head in such a manner that the area is brought up uniformly to the required temperature and held for the specified time. The soak band shall include the nozzle or welded attachment. The soak band shall include a circle that extends beyond the edges of the attachment weld in all directions by a minimum of t or 2 in., whichever is less. The portion of the vessel outside of the soak band shall be protected so that the temperature gradient is not harmful.

(8) Heating of other configurations. Local area heating of other configurations not addressed in (a)(1) through (a)(7) above is permitted, provided that other measures (based upon sufficiently similar, documented experience or evaluation) are taken that consider the effect of thermal gradients, all significant structural discontinuities (such as nozzles, attachments, head-to-shell junctures), and

any mechanical loads which may be present during PWHT. The portion of the vessel or component outside the soak band shall be protected so that the temperature gradient is not harmful.

(b) The temperatures and rates of heating and cooling to be used in postweld heat treatment of vessels constructed of materials for which postweld heat treatment may be required are given in UCS 56, UHT 56, UNF 56, and UHA 32.

(c) The minimum temperature for postweld heat treatment given in Tables UCS 56, UHT 56, and UHA 32 and in UNF 56, shall be the minimum temperature of the plate material of the shell or head of any vessel. Where more than one pressure vessel or pressure vessel part are postweld heat treated in one furnace charge, thermocouples shall be placed on vessels at the bottom, center, and top of the charge, or in other zones of possible temperature variation so that the temperature indicated shall be true temperature for all vessels or parts in those zones.

(d) When pressure parts of two different P. number groups are joined by welding, the postweld heat treatment shall be that specified according to either UCS 56 or UHA 32 for the material requiring the higher postweld heat treatment temperature.

(e) Postweld heat treatment, when required, shall be done before the hydrostatic test and after any welded repairs except as permitted by UCS 56(f). A preliminary hydrostatic test to reveal leaks prior to postweld heat treatment is permissible.

(f) The term nominal thickness as used in Tables UCS 56, UCS 56.1, UHA 32, and UHT 56 is the thickness of the welded joint as defined below. For pressure vessels or parts of pressure vessels being postweld heat treated in a furnace charge, it is the greatest weld thickness in any vessel or vessel part which has not previously been postweld heat treated.

(1) When the welded joint connects parts of the same thickness, using a full penetration butt weld, the

nominal thickness is the total depth of the weld exclusive of any permitted weld reinforcement.

(2) For groove welds, the nominal thickness is the depth of the groove.

(3) For fillet welds, the nominal thickness is the throat dimension. If a fillet weld is used in conjunction with a groove weld, the nominal thickness is the depth of the groove or the throat dimension, whichever is greater.

(4) For stud welds, the nominal thickness shall be the diameter of the stud.

(5) When a welded joint connects parts of unequal thicknesses, the nominal thickness shall be the following:

(a) The thinner of two adjacent butt-welded parts including head-to-shell connections

(b) The thickness of the shell or the fillet weld, whichever is greater, in connections to intermediate heads of the type shown in Figure UW 13.1 sketch (f)

(c) The thickness of the shell in connections to tube sheets, flat heads, covers, flanges, or similar constructions

(d) In Figures UW 16.1 and UW 16.2, the thickness of the weld across the nozzle neck or shell or head or reinforcing pad or attachment fillet weld, whichever is greater

(e) The thickness of the nozzle neck at the joint in nozzle neck-to-flange connections

(f) The thickness of the weld at the point of attachment when a nonpressure part is welded to a pressure part

(g) The thickness of the weld in tube-to-tube-sheet connections.

The thickness of the head, shell, nozzle neck, or other parts as used above shall be the wall thickness of the part at the welded joint under consideration. For plate material, the thickness as shown on the material test report or material certificate of compliance before forming may be used, at manufacturer's option, in lieu of measuring the wall thickness at the welded joint.

(6) For repairs, the nominal thickness is the depth of the repair weld.

Index